Computer Supported Cooperative Work

Springer

London
Berlin
Heidelberg
New York
Barcelona
Hong Kong
Milan
Paris
Singapore
Tokyo

A list of out of print titles is available at the end of the book

Barry Brown, Nicola Green
and Richard Harper (Eds)

Wireless World

Social and Interactional Aspects of the Mobile Age

With 14 Figures

 Springer

Barry Brown
HP Labs, Filton Road, Stoke Gifford, Bristol, BS34 8QZ

Nicola Green
Richard Harper
Digital World Research Centre, School of Human Sciences, University of Surrey, Guildford, Surrey, GU2 7XH

Series Editors
Dan Diaper, PhD, MBCS
Head, Department of Computing, School of Design, Engineering and Computing, Bournemouth University, Talbot Campus, Fern Barrow, Poole, Dorset BH12 5BB, UK
Colston Sanger
Shottersley Research Limited, Little Shottersley, Farnham Lane
Haslemere, Surrey GU27 1HA, UK

British Library Cataloguing in Publication Data
A catalogue record for this book is available from the British Library

Library of Congress Cataloging-in-Publication Data
Wireless world: social and interactional aspects of the mobile age / Barry Brown, Nicola Green, and Richard Harper (eds.).
     p. cm. – (Computer supported cooperative work, ISSN 14311496)
     Includes bibliographical references and index.
     ISBN-13: 978-1-85233-477-2   e-ISBN-13: 978-1-4471-0665-4
     DOI: 10.1007/978-1-4471-0665-4
     1. Telephone, Wireless–Social aspects–History. 2. Wireless communication systems–Social aspects–History. I. Brown, Barry, 1972-II. Green, Nicola. III. Harper, Richard, 1960-IV. Series.

HE9713 .W575 2001
303.48'33–dc21   2001042667

ISBN-13: 978-1-85233-477-2 Springer-Verlag London Berlin Heidelberg
a member of BertelsmannSpringer Science+Business Media GmbH
http://www.springer.co.uk

Typesetting: Gray Publishing, Tunbridge Wells, UK.

34/3830-543210 SPIN 10834346

Country of Origin: United States

Springer
the language of science

Thank you for your order! An invoice is being mailed separately.
For more information on our products and for new book ann

Reference#	0081077720

Ship to:

BAKER & TAYLOR BOOKS
501 S GLADIOLUS STREET
MOMENCE SERVICE CENTER

Tel: 8154722444

MOMENCE   IL   609541799

United States

Quantity	ISBN	Title
1	9781852334772	Wireless World : Social and Inter

1   Total quantity enclosed

0081077720

Packing List

20120723

Please call 800-SPRINGER (1-800-777-4643) if assistance is required.
...uncements, please visit us at springeronline.com/springeralerts.

Customer Order Date	07/23/2012	B & T UPS Collect
		PICKUP

Purchase Order: MOM2569432

...ctional Aspects of the Mobile Age	Author/Editor
	Barry Brown; Nicola Green; Richard Harper

US returns to:
Springer Returns Dept
c/o IPS
1210 Ingram Drive
Chambersburg, PA 17202

Canadian returns to:
Springer
c/o Georgetown Terminal Warehouse
34 Armstrong Avenue
Georgetown, Ontario L7G 4R9

Contents

List of Contributors

Barry Brown
Department of Computer Science, University of Glasgow, Scotland

Elizabeth F Churchill
FX Palo Alto Laboratory Inc

Geoff Cooper
Department of Sociology, University of Surrey, England

Diana Gant
Indiana University, USA

Nicola Green
Digital World Research Centre, University of Surrey, England

Richard Harper
Digital World Research Centre, University of Surrey, England

Vincent Helyar
Serco Usability Services, London, England

Sara Kiesler
Carnegie Mellon University, Pittsburgh, USA

Catrine Larsson
Viktoria Institute, Göteborg, Sweden

Eric Laurier
Department of Geography, University of Glasgow, Scotland

Ged M Murtagh
Digital World Research Centre, University of Surrey, England

Kenton O'Hara
The Appliance Studio, Bristol, England

Leysia Palen
University of Colorado, Boulder, USA

Mark Perry
Dept of Information Systems and Computing, Brunel University, Bristol,
England

Tony Salvador
Intel Corporation, USA

Marilyn Salzman
US WEST Advanced Technologies

Abigail Sellen
Hewlett-Packard Research Labs, Bristol, England

John Sherry
Intel Corporation, USA

Anthony M Townsend
Taub Urban Research Center, New York University, USA

Nina Wakeford
Department of Sociology, University of Surrey, England

Alexandra Weilenmann
Viktoria Institute, Göteborg, Sweden

Part 1
Introduction

Chapter 1
Studying the Use of Mobile Technology

Barry Brown, Department of Computer Science, University of Glasgow, Scotland

1.1 Introduction

Predicting the future is a dangerous game. In the early 1980s the respected consultancy firm McKinsey was asked by AT&T to predict the number of mobile phones which would be in use by the turn of the century. McKinsey confidently predicted that the total market worldwide would be around 900,000, taking into account the poor quality of the devices and the substantial cost of mobile phone calls.[1]

At current growth rates around 900,000 people join a mobile phone network *every day*.[2] Nearly anyone who lives in the developed world will have noticed the proliferation of bleeps, noises and rings from mobile phones in the urban environment. Indeed, the chances are that they would themselves own a mobile phone (as do two-thirds of the British population). The "yuppie-bear"[3] is now a familiar part of our lives, both loved and loathed.

While we might treat McKinsey's figures as a wise example of how foolish it is to predict the future, this would ignore that prediction is an essential part of the technology business. Those who design, sell and market technology must all make predictions about what products or features they expect to be popular, to sell in the marketplace, or to become affordable. These decisions do not just affect new technology enthusiasts; vast sums of money change hands, such as the £22 billion ($35.4bn) paid in the British third-generation (3G) cellular radio frequency auction. In justification of these figures we can read debatable claims that we will soon be shopping on our mobile phones, or that mobile internet use will overtake PC-based internet use (Kanellos, 2001).

It is perhaps surprising then that little research has been done on the non-technical aspects of mobile technology. After all, it is largely the social and cultural aspects which will determine the success or otherwise of these massive investments. For mobile technologies, and especially mobile telephones, are as much social objects as technological ones. They impact how we organise our days and our evenings, how we work, and even how we make new friends. Public places now contain private conversations, text messages disturb intimate moments, and the media hails each new cultural/technological development, from "textual-harassment" to "phone envy". While the technology has certainly changed our culture, culture itself has remade this technology in a thousand different ways.

It was to investigate and explore this "remaking" that we organised the "wireless world" workshop to look at these new technologies. This book brings together the papers presented at that workshop, looking at the social, cultural and interactional aspects of mobile technologies, with contributions from across the social sciences. The focus of this book is on the meanings which people give to their mobile technology, how they integrate them into their work and home lives, and how they interact with those devices, and other people through those devices. This broad agenda means this book contains chapters with very different methodological and theoretical positions. Yet all the chapters share a common concern for understanding mobile technologies in the sites where they are actually used. This includes observations of schoolchildren sharing their mobile phones, the body movements of those using mobile phones, and even data from those who leave their mobile phones switched off. This data means that this book is not just speculation on the culture of the mobile phone, or dramatic predictions of future technologies: it is based on detailed examination of how mobile technologies are actually used now. Neither have we limited this book to voice communications – while the mobile *phone* is the focus here, as is fitting for such a fast moving area, a number of the chapters discuss other mobile technologies and the opportunities and dangers of these new devices.

In editing this book we have attempted to address two different audiences who do not normally converse. With a technology which is so widespread, and perhaps strangely representative of its age, there is something to say to social researchers from understanding its use. As Cooper puts it in the next chapter, while there is a sense that mobile phones are hard to track down, or "easy to just miss", these devices do impact and change our social interactions. We might dismiss mobile phones as trivial little devices – but it can be argued that these devices have something wider to say about society, and even changes in society itself. Perhaps by dismissing the mobile phone we are dismissing the very stuff of society. As Latour argues, are objects such as the mobile phone "the missing masses" from the social sciences? (Latour, 1992, 2000). After all, much of our lives involve production, consumption, or use of objects. To take another "trivial" example, household technologies like the microwave have certainly played a role in gender divisions (Cowan, 1983). It is not that gender comes from the microwave, just as a society-destroying acid does not seep from new technologies, but rather that these devices come to be used to both replicate and change the everyday ways in which our lives are lived. For social science then, the object can be a hook to understanding a little of individuals' lives and concerns – particularly in times of societal change. As Dent puts it:

> Sociology needs to begin to attend to the ways in which interaction with objects is part and parcel with the social interaction which gives rise to social forms. We express ourselves as part of *this* society through the ways we live with and use objects. (Dant, 1999: 2)

This concern for talking to social science is also combined with an attempt to consider what we can say for the design of technology. While this may seem an unusual combination, as we will discuss later, it has its roots in much recent work studying technology in use, particularly in the computer supported collaborative work (CSCW) field. We would argue that if we can gain an understanding of the use of a technology, then we should not be afraid of moves towards informing design. If we do not, then technological decisions, and therefore technological predictions, will be based on the opinions of those who know the technology intimately, yet know little of their use.

1.2 Understanding the Technology

We will return to these points after discussing something of the technology itself. For the mobile phone has been described by some writers as a pivotal technological device, that represents more than just a new technology but something of a new *type* of technology. Mobile phones, unlike personal computers, are small, mobile, constantly on, and potentially constantly connected. It is these features of the device which have led to a great amount of interest (and investment) and the claim that mobile phones are the first of a wave of these mobile technologies (Goodin and Grice, 1998). Much of this interest results from the arguments that the mobile phone will overtake the PC as the technological device of choice. This argument can be traced to the "disappearing computer" argument made by Weiser and expanded on by Norman (Norman, 1998; Weiser, 1991).

Weiser and Norman argue that the personal computer is something of a successful failure. In an attempt to be a multifunction device the PC has ended up being too complicated and overburdened by the need to support a huge range of different users and tasks. This complexity means that for most individuals the PC is just too confusing for what they want to do, and thus inevitably frustrates its users. Despite its huge popularity most PC users have little idea of how to use anything but the simplest of its features. As a solution, Norman and Weiser argue that the general purpose PC will disappear out of sight, to be replaced by single function appliances. Just as electrical engines moved from being devices in themselves and disappeared into other devices, so the computer will "disappear", and become a ubiquitous unnoticed feature of the environment. The mobile phone clearly demonstrates this argument; we forget that inside most phones is enough computer power to match the average home computer of the 1980s. The computer has "disappeared" into the mobile phone, a single function device, in a form which means we do not even notice the computer inside the phone.

Certainly these arguments have some merit; there is a role for simpler, easier-to-use technological devices. However, there are some economic barriers to this happening. By the simple fact of mass production, the price of electronic devices is generally determined by the size of each production run. Since so many different applications can be run on the PC, massive numbers of PCs are built at a relatively low cost for a wide range of different applications. This means that single function (and thus smaller market) devices currently suffer from high cost. For example, wireless "web-tablets" which allow surfing the web from a non-PC appliance are considerably more expensive than a PC which has more functionality. This cost places a number of economic barriers to the greater success of single function appliances.

There is, however, a more fundamental problem (Brown, 2000). There are many different successful devices on the market which are in fact multi, rather than single function. The video recorder, for example, plays back pre-recorded videos as well as time-shifting television programs. It is perhaps then not simply single-function devices we should look for, but rather a manageable collection of functionality. Indeed, if we consider the mobile phone, it includes a wide range of different features and technologies, all of which have contributed to its success (Table 1.1).

While it is easy to focus on the first function in the list in Table 1.1, it is this complete package which has been important. Together all these different features have contributed to the mobile phone's success. Take, for example, the last two features on the list. For many consumers mobile phones will have been their first experience of caller ID and voice mail technology.

In one organisation we have studied, these technologies were combined in the use of a conventional fixed line phone system to screen-calls (Brown and Perry, 2000). This organisation has installed a fairly advanced (for its time) fixed line telephone with both caller ID and voice mail functionality. Together these inventions changed how the phone was answered in that organisation. Rather than the usual procedure of answering the phone, the receiver of the call would turn, glance at his or her phone, either answer the phone or push a button to redirect the call onto voice mail. One interesting example of this happened when we were interviewing an accounts clerk. The phone rang during the interview and the clerk glanced at the phone and then ignored the call. When the phone rang again for a second time almost immediately afterwards, she picked up the handset: "hi ... yeah, ... I saw that you'd called twice, so I thought I better answer it."

In ignoring the call, the account clerk had checked to see the number of the caller. On the second call, she recognised that the caller had rung again and ascertained that it must be about an important matter. This "glance then ignore" procedure appeared to involve checking the caller's number as a clue to the importance of the message, and then making a decision about whether it is important enough to answer, or leave to the voice mail. Interestingly, the caller was also aware of this behaviour – she calls twice, knowing that she may have been actively ignored on her first attempt.

Whilst this combination of caller ID and voice mail was presumably accidental, together they provided excellent support for not answering the phone. Voice Mail abates the rudeness of not answering a call by offering a facility to continue the communication asynchronously, without a direct connection. caller ID, on the other hand, provides call recipients with information to support *the option* of

Table 1.1. Eight different features of the modern mobile phone

- Send and receive telephone calls from/to land lines
- Send and receive telephone calls from other mobile phones
- Telephone book and limited calendar functionality
- Mobile short messaging of text, graphics and music (send and receive)
- Alarm clock, calculator, games
- Online data access through WAP or IMODE services
- Caller identification with name recognition from telephone book
- Personal voice mail with mobile access

answering the phone, rather than a compulsion to do so. With caller ID, important callers (or those who could not be dealt with through voice mail) could be answered immediately, whilst passing calls off which could be dealt with at a more convenient time. So while the mobility of the mobile phone facilitates an *increased* level of connection and communication, these two technologies support more *appropriate* communication. These technologies were appropriated by their users to support not answering a ringing phone – an important facility for busy office staff when they did not want to be disturbed.

With a mobile phone this functionality becomes even more important, since (as Gant and Keisler discuss in Chapter 9) the range of environments where we can receive calls is greatly increased. Indeed a careful mobile phone caller can even detect the different number of rings on calling a mobile phone, depending on whether the phone is switched off, or the call has been screened.

1.3 Tracing the Non-development of the Mobile Phone

While these different features contribute to the popularity of the phone it is of course the mobile component of the *mobile* phone which is its most important feature. We do not buy our mobile phones because they come with voice mail, or for the games you can play on the phone. Rather, it is the ability to be in contact while we are outside the reach of conventional land phones.

Before we discuss the use of mobile phones it is useful to consider briefly the history of this invention. Indeed, the development of the mobile phone is of interest in itself, since it is almost a history of *non*-development. During the cold war money was poured into research efforts such as artificial intelligence, computer graphics and nuclear weapons, whereas the mobile phone struggled to gain investment and effort (Edwards, 1997). The first commercial mobile phone systems were up and running in the 1940s, yet it took over 30 years to develop a mass market mobile phone system. Looking at the early history of the mobile phone gives an example of how a technology can be delayed by decisions to favour other technologies – the technology predictions that we mentioned in the introduction.

While the first mobile telephone systems were commercially available in the 1940s, these systems were based around fairly crude technology. Car-based radios would broadcast and receive transmissions from a single fixed base station, where the radio channel would be connected to a land phone line. To use the system a user would manually search to find a free radio frequency to use, pick up the receiver and be connected on that radio frequency to an operator at the base station. They would then quote their subscriber number (for billing purposes), ask for a particular telephone number, and the operator would then dial the call and link the land telephone line with the radio channel (Douglas, 1964; Peterson, 1947).

Although this system was straightforward it suffered from a chronic lack of capacity. Since radio transmissions can travel a considerable distance, the frequency being used for a call cannot be reused. It is blocked by that call for as far as the radio transmissions can be received. This means that there has to be a separate channel for each call. This lack of channels effectively crippled this early mobile phone system, and demand constantly outstripped supply. By 1976 in the USA, for example, while 44,000 people had mobile phones there were 20,000 individuals sitting on 5–10-year waiting lists (Roessner *et al.*, 1998). So although very popular, the lack of frequency essentially prevented the early mobile phone systems from becoming mass market devices.[4]

Surprisingly, a solution to the capacity problem already existed at the time of the development of these systems. The solution was developed in Bell Labs, AT&T's massive research lab, in 1947 and was based around splitting the geographic coverage area into individual "cells" (*ibid.*) Within each cell radio transmissions are broadcast at such a strength that the transmissions will not reach adjacent cells.[5] This means that frequencies for calls can be reused between cells. Moreover, if you need more capacity, you can cut the power of the transmitters and split the cell up into four smaller cells, each with the capacity of the previous larger cell.

It was this invention which made *cellular* phones possible, and in turn mass-market mobile phones. By reusing frequencies from one cell to another, call capacity could be greatly increased. However this technology presented a number of technological challenges. The mobile phones must be able to maintain a call as the user moves from one cell to another, by handing over the call from one base station to another. The phone itself must also be compact enough to be mobile – at least small enough to fit inside a car. The phone must also be able to transmit on a narrow frequency band, without interfering with other calls. These problems, however, were not insurmountable, and had all been solved by the late 1960s. In 1969 the first non-experimental cellular call was made from a service running on Metroliner trains from New York City to Washington DC, where passengers found that they could "conveniently make telephone calls while racing along at better than 100 miles an hour" (Paul, 1969). This also means that (ironically) the first cellular phone calls were made from a payphone. By 1973 the cellular technology was even small enough to be handheld – Dr Martin Cooper, a project manager at Motorola, made a call on a portable cell phone from a New York street near the Manhattan Hilton to his rival, Joel Engel, head of research at Bell Labs (Farley, 2000).

Despite these developments, mobile phones were delayed throughout the 1970s. This further delay was caused not so much by technology, but by the regulatory and business decisions made by the government and phone companies. In particular, the FCC (who control the allocation of radio spectrum in the USA) were hesitant to allocate spectrum to such an unproven technology, over frequency for new television channels. There were also lengthy legal disputes between the operators of the existing non-cellular phone services and AT&T which delayed services for much of the 1970s.

By the time the first mass-market USA commercial cellular phone system started in 1983, 37 years had passed since the first carphone service. It is tempting to blame the government, in the form of the FCC, for the huge delay in the development of cellular phones – and they certainly had a major role. However, these regulatory constraints did not exist in other countries, yet these countries introduced cell phones only marginally quicker. For example, the Scandinavian cellular system was launched in 1981 when Sweden, Norway, Denmark and Finland adopted the joint NMT400 (Nordic Mobile Telephone) system, only two years ahead of the USA (*ibid.*).

A more fundamental delay in developing the mobile telephone came from the hesitancy of researchers to do cellular telephones research. Throughout the 1960s and 1970s academic research on the problems of cellular was rare. Working in the area, as one academic put it was like being "lost in the desert". Another academic researcher described the area as "grubby" (Roessner, *et al.*, 1998; 17). Even within Bell Labs, where most of the work on cellular communications was done, the situation was similar. Before the 1980s, cellular researchers constituted a tiny fraction of Bell Labs' thousands of scientists and engineers, and this area was considered by many to be a professional backwater (*ibid.*: 18).

Unfortunately, it is impossible to prove that a technology *could* have developed faster with more attention. Certainly, it is possible that other factors could explain the rather pedestrian pace of cellular phone development. However, a short comparison can be made with the efforts over the years to design and build videophone systems. During the 1960s and 1970s, video phones appeared to have considerably more appeal and attention for those developing technology than mobile phones. Indeed, AT&T (which at that time essentially controlled the US phone market) spent over $500 million during the 1970s to develop and market the videophone (Nole, 1992). Yet, in market trials the response was extremely limited. For example, AT&T launched a public phone system in April 1971 in Chicago. By 1973 there were only 100 public videophone subscribers. In the mid-1970s, AT&T then attempted to sell their system into the teleconferencing market. Again, despite giving free calls to the executives of major firms – using a very sophisticated conferencing system with multiple screens – few came back to use the system. An internal AT&T corporate system likewise experienced low usage. Despite high level management commitment to the service, AT&T failed to find sufficient customers for the videophone in various market trials and studies throughout the 1970s, both in the home and in the office.

As Michael Noll, the designer of the video phone booth used in the movie *2001: A Space Odyssey* who was personally involved in the AT&T trials, puts it:

> It is tempting indeed to become overly fascinated with technology. I remember that I too was swept along with enthusiasm for the picture phone. [...] but customers and their response are the ultimate determinant of the success and failure of new products and services – and most consumers simply did not want picture phone service (*ibid.*: 316)

Despite these failures, companies today are still attempting to develop videophones. In 1987, Mitsubishi introduced its VisiTel Visual telephone system – it failed in the marketplace. AT&T – again – attempted to market a visual telephone in 1992, which failed (*ibid.*). Orange (a mobile phone network operator) are also currently attempting to market a portable videophone. It appears that again and again, despite the many failures, technology organisations seek to design and build videophone systems. It appears to have some sort of strange attraction – perhaps even something of a *pathology* in the mind of technology developers. However, cellular technology systems seemed to have no similar attraction. Throughout its

30 years of development it was considered a backwater, with little research and attention. Perhaps then the cell phone is more an example of *non*-development, as other research areas took attention and funds.

It is tempting to suggest why the cellular phone has had less attraction than the videophone to technology developers. However, as the McKinsey prediction in the introduction to this chapter shows, the rise and success of the cell phone was not easily predictable. However predictions of the future are implicit in technology research and some potential inventions must be favoured over others – in this case the unsuccessful over the successful. While it is perhaps overly optimistic to suggest that social researchers might have been able to change these events, this at least suggests a role for those who study the *use* of technology. If one has to make decisions about the success or failure of technology, detailed descriptions of its use could improve the quality of these decisions.

As an example, in the conclusion to this book Harper discusses some of the attention which has been given to WAP technology, and the ability to access information, and even go shopping (so called m-commerce) from the mobile phone. Yet a number of the chapters in this book emphasise how the value of the technology is in enabling communications between individuals rather than between individuals and institutions. This suggests that there may be as much potential in allowing increased exchange of personal data between individuals – such as swapping games, applications, photos, documents, sound samples and so on between phones. While studies of use can never help us to make *definite* predictions we would argue that it can certainly help technologists to make *better* predictions.

1.4 Understanding the Technology in Use

After this short history we can now go on to consider some of the research fields from which the work in this book is presented. As mentioned in the introduction, this book spans two very different research fields. The first group of work originates in sociology, specifically in the field of social studies of science and technology (SST). While SST is a large and growing field with arguments that cannot easily be presented briefly, at its heart there is a concern for how technology and science are interconnected, and in turn are changing society. *Techno-science*, the amalgamation of technology and science, affects what we eat, how we communicate, what medicines we take – now even how we reproduce. It is this "techno-science monster" which has been SST's main subject (Fuller, 1997; MacKenzie and Wajcman, 1985). While little work in SST specifically looks at mobile technologies, as Chapters 2–5 in this book show, SST has a number of valuable lessons for those of us wanting to understand mobility and mobile devices. Perhaps most importantly SST makes our assumptions about these devices problematic. For example, while it might seem simple to see mobile phones as private devices, in competition with public payphones, mobile phones are also public devices in that they are used in public spaces. Indeed, to what extent can spaces be easily categorised as public or private? Work in SST also problemises a clear divide between humans and technology, since a clear divide ignores the complex interactions between objects and humans in a technical system. When looking at complex systems it is as important to discuss how the objects act – how they have agency, as well as the humans. When our mobile phone keeps ringing because we have a voice mail message we don't want to listen to, does not the mobile phone system have some agency of its own?

In contrast, the second field of origin for the work in this book is the more technology focused fields of human computer interaction field (HCI), and computer supported collaborative work field (CSCW). Within this book this influence can be seen in Chapters 6, 7, 8, 10 and 12. Within these fields there has been an increasing recognition of the value of studying technology in use to inform the development of new technology (Hughes *et al.*, 1994). In particular, many of these studies have been based around ethnographic observations and video recordings of work practice (Heath *et al.*, 2000). These studies use empirical studies to produce analysis specifically for the purposes of design. The focus here has been on the everyday practices which are involved in carrying out work (and to an increasing extent, leisure). Perhaps somewhat surprisingly these studies are somewhat at odds with the traditional sociology of work. For example, even in a relatively progressive text on the subject, such as Grint's review (Grint, 1991), there is a lack of description of anyone *doing* anything. The topics discussed are race, ethnicity, patriarchy, trade unions, class, industrial conflict, organisational cultures and modern capitalism. There is no room for even a brief discussion on how work as an activity is organised. This criticism of sociology has motivated much of the social research in HCI and CSCW. As Orr puts it:

> This is the main problem with all this literature. It is not well grounded in analysis of work practice, so its presumptions and prescriptions of what is to be done are not based on what is done and what needs to be done, on the reality of the job, the task to be accomplished. (Orr, 1996: 151)

Increasingly these methods are being transferred to fields outside work environments, such as in the home (Hughes *et al.*, 2000; O'Brien and Rodden, 1997), or in leisure activates such as creating or listening to music (Bowers and Hellstrom, 2000). In this book these concerns have been developed into looking specifically at mobile devices, and what we might learn from these technologies in use.

1.5 Overview of the Book

This book then is a combination of these two approaches around the shared topic of mobile technology. The book is divided into three parts, to indicate the diverse viewpoints and motivations behind the different chapters. In the first part, "Locating mobile technology", the focus is on how we might understand the technology and its effect on both us and our interactions. In the second part, "From ethnography to use", the focus is on empirical description of the use of these new mobile technologies. In the final part, "From use to design", we take these lessons back to design through a series of studies where the results are used to inform the design of new mobile technologies.

1.5.1 Situating Mobile Technology

This first section locates mobile technology with reference to current debates in social theory and specifically, science and technology studies. In Chapter 2 Geoff Cooper starts the book by situating our investigations in the context of social theory. As he shows, some of the qualities of this technology can be obscured by its status as both an already routine feature of everyday life, and yet one which continues to be subject to rapid change. Initial distinctions such as that between the private and the personal might seem initially relevant to mobile technologies; yet these are problematic under further investigation.

Following on this theme, in Chapter 3 Nicola Green discusses the use of mobile technologies for surveillance and control. In asking "who is watching whom?" Green emphasises that these technologies are as much about individuals monitoring each other, as organisations or the state monitoring us. Surveillance and control are also highly contestable phenomena. For example, when someone asks where you are, *you can always lie.*

Or perhaps more importantly, one can give an ambiguous answer. As Eric Laurier shows in Chapter 4, this control of *location* is a defining aspect of the use of mobile phones. Location is something which is manipulated so as to produce spaces which make work possible – be that with a hands – free kit in a car, or through a conversation in a local café. Laurier's discussion of mobile workers using their cars as mobile offices shows how the mobile phone is used as a tool to reinvent space, and reclaim "dead spaces" for work and interaction.

In Chapter 5, the final chapter in this section, Anthony Townsend takes a step outward and asks what influence mobile technology has on the concept of the city. He argues that the mobile phone has resulted in a quickening of the pace of urban life at an aggregate level. This results in a dramatic increase in the *metabolism* of urban systems. This increase in the metabolism is something that a new decentralised city plan would have to take into account, as processes in the city can change non-linearly and beyond simple geographic co-location.

1.5.2 From Ethnography to Use

In the second section these discussions are taken in a more strongly empirical direction, with a focus specifically on investigating the practices of mobile technologies use. In this section there are four chapters which are based around studies of the technology and its use *in situ.*

In Chapter 6, Ged Murtagh looks at some of the ordinary features of mobile phone use in train carriages. Taking an ethnomethodological approach, his chapter explores how individuals define the parameters of appropriate phone use within this setting. Ethnomethodology is an approach which has been popular in the study of technology use for design, specifically in studies of the workplace. In this chapter, a similar approach is taken which focuses on the unfolding details of particular situations, and in particular the importance of bodily movement and eye contact.

Chapter 7, by Alexandra Weilenmann and Catrine Larsson, takes a similar ethnomethodological perspective, but in this case they look at how teenagers use mobile phones, and in particular how they are shared in public places. For rather than phones being a personal device, Weilenmann and Larsson show how phones operate as a shared resource, both for callers and the called. A simple notion of the phone as an individual device appears problematic when one considers how much the teenagers they studied shared their phones and also how much the usage of mobile phones was a public event to be used in interaction with others locally present.

In Chapter 8, John Sherry and Tony Salvador share a common concern for understanding the details of mobile technology use, although they move the investigation onto the work environment. Using the metaphor of jazz they explore the different balances involved in mobile work. Quoting Charles Mingus: "A pure genius of jazz is manifested when he and the rest of the orchestra run around the room while the rhythm section grimaces and dances around their instruments." As they show, an important aspect of mobile work is maintaining order when the world collapses around you – be that through cancelled trains or broken cars. Mobile technologies are one tool for balancing activity as plans and situations change and are rearranged.

This concern for balance is also seen in Chapter 9 by Diana Gant and Sara Kiesler. Their chapter explores how the line between personal and work life is blurred by the use of mobile technologies. In particular, the mobile phone encourages a form of flexibility which causes work and social time to become more mixed. This causes problems as existing boundaries break down and are not replaced.

1.5.3 From Use to Design

As well as understanding practice, it was an important aim of this collection to inform design, and in the final section of the book the chapters address specifically how we can advise and inform technology design from studies of use.

In Chapter 10, Leysia Palen and Marilyn Salzman demonstrate this. Their chapter describes a study of new users of mobile phones and the confusions and problems they experience. Using this data they separate mobile phone technology into *hardware*, *software*, *netware* and *bizware*. These distinctions come not strictly from the technology but from the interaction between usage and technology. For each of these different aspects they draw implications for improving the design of current technology and services.

In a similar way in Chapter 11, Elizabeth Churchill and Nina Wakeford use empirical data to develop a set of *design dimensions* for mobile technology – from tight mobility to loose mobility, and close information to far information. These concepts have helped designers at Fuji-Xerox to design mobile technologies which fit different forms of collaboration for different settings. In this way the aim is to design technology which fits particular interaction situations, rather than the interaction being forced into a particular technology.

In the next chapter, Chapter 12, Kenton O'Hara *et al.* discuss the use of documents by mobile workers. Their empirical work uncovers the ways in which documents are an essential part of working with mobile phones. Again, this leads on to design where the authors consider how mobile document technologies may be designed so as to better support this mobile document use.

In the final chapter in this section, Chapter 13, our concern turns to the usability of a specific mobile technology: the WAP (wireless application protocol) phone. Vincent Heylar's chapter discusses the problems and confusions of this technology using data from usability tests of WAP phones. As he concludes, research can help to ensure that lessons are learnt from the current generations of mobile technology, to inform the design of the next generation.

As can be seen by these short summaries, this volume covers a range of different materials from theory, to use, to design. But at the heart of these investigations is a concern that we do not lose sight of the topic under study. For while it is *mobility* that is the most important aspect of these new technologies, it is mobility as an *empirical* fact. It is mobility in terms of attempts to make use of time while waiting for public transport, or through relaying news to those not at home. These are examples of the lived experience of mobility – how individuals and groups use mobility for their own purposes or how they must work in the envelope which mobility provides. It is this topic above all others which motivates the chapters in this book.

Notes

1. This story was reported in *The Economist* (Anonymous, 1999).
2. Around 400 million mobile phones were sold in 2000, although many of these were replacement handsets (Charny, 2001).
3. As it is known in Sweden (Kopomaa, 2000: 37).
4. While there is little written about the early users of these systems, one observation (Farley, 2000) concerns the ease of defrauding these early systems. With early systems, when a user placed a call they would verbally give their subscriber number, offering the opportunity to give someone else's number and have your charges go elsewhere. Even when (in later systems) the subscriber number was passed automatically by the radio transmitter, the radio transmitter could be rewired so as to change the number it gave. It seems that then, as now, fraud progressed along with new technological developments.
5. Actually adjacent-but-one cells.

References

Anonymous (1999) Cutting the cord. *The Economist*. Oct 7, 1999. Available on the internet at: http://www.economist.com/displayStory.cfm?Story_ID=236152.

Bowers J and Hellstrom SO (2000) Simple interfaces to complex sound in improvised music. In *CHI'2000 extended abstracts*. The Hague: ACM Press.

Brown B and Perry M (2000) Why don't telephones have off switches? Understanding the use of everyday technologies. *Interacting with Computers*, 12 (6), pp. 623–634.

Brown B (2000) The future of the personal computer in the home. *Personal Technologies*, 4 (1), pp. 39–44.

Charny B (2001) Nokia distances itself from the pack. *CNET News.* Available on the internet at: http://news.cnet.com/news/0-1004-201-4826860-0.html?tag=mn_hd.

Cowan RS (1983) *More work for mother: The ironies of household technology from the open hearth to the microwave.* New York: Basic Books.

Dant T (1999) *Material culture in the social world.* Buckingham: Open University Press.

Douglas VA (1964) The MJ mobile radio telephone system. *Bell Laboratories Record,* p. 383.

Edwards PN (1997) *The closed world: Computers and the politics of discourse in cold war America.* Boston, MA: MIT Press.

Farley T (2000) Mobile phone history. *Private Line.* A website of enquiry into the telephone system. Available on the internet at: www.privateline.com.

Fuller S (1997) *Science.* Oxford: Oxford University Press.

Goodin D and Grice C (1998) Are PDAs the wave of the future? *CNET News.* July 14, 1998. Available on the internet at: http://news.cnet.com/news/0-1003-200-331207.html.

Grint K (1991) *The sociology of work: an introduction.* Oxford: Polity.

Heath C, Knoblauch H and Luff P (2000) Technology and social interaction: The emergence of "workplace studies". *British Journal of Sociology,* 51 (2), pp. 299–320.

Hughes J, O'Brien J, Rodden T, Roucefield M and Viller S (2000) Patterns of home life: Informing design for domestic environments. *Personal Technologies,* 4 pp. 25–38.

Hughes JA, King V, Rodden T and Andersen H (1994) Moving out of the control room: Ethnography in system design. In *Proceedings of CSCW '94,* Chapel Hill, North Carolina.

Kanellos M (2001) I-mode cell phones could rival PCs, exec says. *CNET News,* Available on the internet: http://news.cnet.com/news/0-1004-200-4720953.html.

Kopomaa T (2000) *The city in your pocket.* Helsinki: Gaudamus.

Latour B (1992) The sociology of a few mundane artefacts. In Bijker WE and Law J (Eds) *Shaping technology/building society.* Cambridge, MA: MIT Press.

Latour B (2000) When things strike back: A possible contribution of "science studies" to the social sciences. *British Journal of Sociology,* 51 (1), pp. 107–124.

MacKenzie D and Wajcman J (1985) *The social shaping of technology.* Milton Keynes: Open University Press.

Nole AM (1992) Anatomy of a failure: picturephone revisited. *Telecommunications policy,* May/June, pp. 307–316.

Norman D (1998) *The invisible computer.* Cambridge, MA: MIT Press.

O'Brien J and Rodden T (1997) Interactive systems in domestic environments. In *Proceedings of the ACM conference on designing interactive systems* – DIS '97: ACM Press.

Orr JE (1996) *Talking about machines: An ethnography of a modern job.* Ithaca, NY: ILR Press.

Paul CE (1969) Telephones aboard the metroliner. *Bell Laboratories Record,* p. 77, March 1969.

Peterson AC (1947) Vehicle radiotelephony becomes a bell system practice. *Bell Laboratories Record,* p. 137, April 1947.

Roessner D, Carr R, Feller I, McGeary M and Newman N (1998) The role of the NSF's support of engineering in enabling technological innovation: *Phase II final report to the NSF.* Arlington, VA., SRI International. Avaliable on the internet at: http://www.sri.com/policy/stp/techin2/chp4.html.

Weiser M (1991) The computer for the 21st century. *Scientific American,* 265 (3), pp. 94–104.

Part 2
Locating Mobile Technology

Chapter 2
The Mutable Mobile: Social Theory in the Wireless World

Geoff Cooper, Department of Sociology, University of Surrey, England

2.1 Introduction

> We cannot say anymore that the immutable is truth, and that the mobile, transitory is appearance. Adorno (1973: 361)

This chapter attempts to bring an empirical phenomenon, the mobile, into some kind of theoretical focus: that is to say, to begin to set out some aspects of its possible sociological significance. The technology and the behaviour it facilitates are already ubiquitous, and the wealth of folk-lore that surrounds it indicates that it is certainly a note-worthy social phenomenon. At the same time, its ready-to-hand (or even hands-free) quality, and the very speed of its development and adoption suggest that there is some value in considering how it can be rendered visible; for some of its qualities might be said to be obscured by its status as both an already routine feature of everyday life, and yet one which continues to be subject to rapid change.

The principal aim then is neither to locate the mobile and its use within some grand theoretical narrative of the state of communication and society, nor to survey a range of possibly relevant social theories but, more modestly, to consider some theoretical resources (from within sociology, science and technology studies, philosophy and media studies) for thinking about certain features of a technology which seems manifestly socially significant – it introduces changes in social interaction for example – but, in some respects, evades sociological analysis. It should be noted that the features of interest emerge from an ongoing empirical research project, and that part of my argument will be that theoretical work carried out in isolation from the study of the practicalities of situated mobile use can easily go astray.[1]

The title is an allusion to Latour's (1986) concept of an "immutable mobile" which, playfully reconciling two traditionally opposed elements within philosophical thought, denotes a technology (an inscription or representation such as a map for example) in which the portability of unchangeable (though recombinable) information from one setting to another makes possible action at a distance. The mobile on which we are focusing has something in common with this but, with respect to emerging patterns of use and social behaviour, can also be said to be mutable, transitory and malleable.

The chapter is organised as follows. I first consider the issue of the technology's transparency, and the challenge that this poses to analysis. I then look at three features of the mobile and its use: the conjunction of remote and co-present inter-action; mobility and location; and being available to others. My argument through-out is that one reason why the mobile is of interest is because of its strategic and reflexive value for occasioning the reconsideration of theoretical categories and modes of analysis.

2.2 Transparency, Visibility, Audibility

The mobile is in many ways an elusive phenomenon to conceptualise. It seems to belong to the category of "new media" but much of that literature is not pertinent; for the mobile, resembling in part its ancestor the fixed-line phone, is relatively transparent, at least at an intuitive level. Similarly, terms such as "virtual" or "cyberspace" do not seem to fit, at least not yet. Standing at the threshold of "con-vergence" or its integration with different media – whatever form that may take – complicates matters further: for example, a convergence of technologies also means a convergence of literatures, with implications for what might count for analytic competence. When we consider the usefulness of specific existing litera-tures, we find similar problems. The sociological literature contains a number of terms which seem apt, but have had somewhat different referents: social mobility, the problematising of the public/private distinction, or the structural transforma-tion of the public sphere (Habermas, 1989) for example. They are relevant con-cepts, but to "apply" them to the study of the mobile is, often, to give them an interestingly literal twist.

These factors contribute to an uncertainty about the analytic specificity of the mobile. Is it simply an instance of wider social and technological change, or does it have greater moment than this? Relatedly, what is the most profitable level or unit of analysis? Do the kinds of distinctions – macro, micro and so forth – that are sometimes made have any value in this domain, or is the mobile better understood as something that makes possible the movement from one level to another?[2] Such questions are not unique to mobiles but they add to the sense that the mobile is hard to track down, easy to just miss. These are analytic problems but also part of the phenomenon itself.

Although it is the case that the mobile is frequently represented, in advertising for example, as a technology of the future that is, somehow, in our hands today, analysing the mobile necessarily takes us back to the fixed-line telephone (as we now have to call it). This is by now a relatively naturalised technology, the most transparent of electronic media. This is apparent at the phenomenological level when we use the phone; as Meyrowitz comments "speaking to someone on the tele-phone, for example, is so natural that we almost forget about the intervening medium" (1985: 109). It is also apparent at the analytic level: the telephone has attracted relatively little attention in comparison with other media, and it is tempt-ing to say that this is precisely because of its success as being simply a medium, conveyor of talk. Barthes argued, contentiously, that photography was different to other visual media because it was not a form of representation (Barthes, 1993). The argument would seem to apply less contentiously to the phone. Conversation analysis, for example, found its inspiration and point of genesis in Harvey Sacks' study of a corpus of transcribed telephone calls, and the "wild" idea that occurred to him in the course of this (Schegloff, 1992: xvi); yet, while such materials continued to be used in many early CA studies, the topic was conversation *per se*, and relatively little consideration was given to the specificity of the telephone and the formal properties of interaction that it might afford.

In an aside about the phone which is remarkable for its prescience – "people share it, so that there are not yet things where you can call up a particular person and get them, or get nothing" (Sacks, 1992: 548) – Sacks notes one reason for the failure of technocratic dreams with respect to new communication technologies: what in fact happens is that "the object is made at home in the world that has whatever organization it already has" (*ibid.*: 549). This not only provides a useful articulation of the limitations of deterministic views of the impact of technology on society, a well trodden if still indispensable theme in science and technology studies,[3] but also hints at the temporal dimensions of the process by which a technology is made routine, or achieves transparency. This dynamic is more clearly delineated by Marvin, who argues that new communications technologies "are always introduced into a pattern of tension created by the co-existence of old and new" (Marvin, 1988: 8). This suggests that we have or have had an opportunity, albeit it a brief one, to study the period of adjustment and friction that accompanies the introduction of a new communications technology.

This historically specific period of adjustment therefore provides one route to a proper view of the social significance of the mobile, since a background of existing social practice is still discernible and available as a point of contrast. It is notable however that the recent proliferation of research into the mobile can be more directly related to the promise of convergence, and the imminence of its changed status as point of access to other forms of information, and in particular to more widely discussed visually based technologies. The mobile phone, as opposed to the mobile device, remains relatively unremarkable for social science.

The mobile may also be said to facilitate the transparency of the world, for it promises to make possible communication at any time from any place, to eradicate communication-free pockets. (The film *Blair Witch Project* already looks anachronistic to this viewer, since the central theme of the unlikely presence of the archaic and mysterious just off the American beaten track would collapse had one of the three protagonists had a mobile.[4]) However, as we shall see, the significance of this facility can be viewed in very different ways.

As a phenomenon, the mobile can be approached in a number of ways, formulated according to different scales. I want to begin at ground level and look at the conjunction of two forms of interaction which constitute a routine feature of its use.

2.3 Conjunctions of Remote and Co-present Interaction

Simmel, writing a century ago about the shock of the experience of modernity in the metropolis, commented thus:

> Interpersonal relationships in big cities are distinguished by a marked prepon-
> derance of the activity of the eye over the activity of the ear. The main reason for
> this is the public means of transportation. Before the development of buses, rail-
> roads, and trams in the nineteenth century, people had never been in a position of
> having to look at one another for long minutes or even hours without speaking to
> one another. Simmel (cited in Benjamin, 1983: 38)

It is worth briefly considering whether the use of the mobile in public places, including the confined ones that Simmel describes, does not accentuate this experience or take it a stage further. To begin an exploration of this issue, I make use of a distinction between public and private; however, I shall go on to argue that these categories can only take us so far, and that the mobile can provide one approach – there are many others – towards their reformulation.[5]

Simmel's analysis of the metropolitan experience places emphasis on the conflict of private – conceptualised, problematically from the perspective of a century later, as inner subjectivity – and public. He suggests for example that one way in which the individual can deal with the shocks of modernity is through the securing of "an island of subjectivity, a secret, closed-off sphere of privacy" (Frisby, 1985: 105); thus, the individual can handle the apparent horrors of inflicted co-presence and other disorientating phenomena. The predominance of the eye over the ear is manifested in the possibility of, or preference for, silence. There is a strong sense here that the emerging form of public life is based on an aggregation of individuals with their own private, possibly incompatible lives; Benjamin makes a similar point, arguing that public gatherings of certain kinds "have only a statistical existence" and are "abstract" even "monstrous" since individuals' interests (including class interests) are radically different (*ibid.*: 250).

The use of the mobile in certain public spaces makes the relation of private and public slightly different. No longer is the private conceivable as what goes on, discreetly, in the life of the individual away from the public domain, or as subsequently represented in individual consciousness; furthermore, although it is still the case that the co-present tend not to speak to each other, they *can* now have conversations with remote others which are (half) audible to all. The co-existence of, and potential friction between public and private are now material and observable phenomena.

Goffman (1971) can be seen as developing Simmel's interest in public behaviour and its normative dimensions, and his work provides some useful resources for thinking about mobile use in urban public space.[6] First, he notes that in certain situations it is customary not only to not speak to others but to avoid looking directly at others: "civil inattention". The management of gaze can thus be regarded as one of the ways in which the boundary between public and private is negotiated (see Chapter 6). Goffman notes other normative features of public behaviour which can connote a kind of respect for the other, such as avoiding walking in between two people who are in conversation. We can however think of situations in which the normative parameters are less well established than in Goffman's examples – is it all right to walk between two people when one is taking a photograph of the other? – and the use of the mobile in public arguably creates such a situation. There is both uncertainty and perhaps, in Marvin's (1988) terms, a tension between established patterns and norms of social behaviour and those facilitated by the new technology.[7]

A second concept of Goffman's, the "tie-sign", is useful here. These are forms of behaviour, for instance gestures, which display social affiliation. Goffman notes that when speaking on the phone in the presence of a known other, the speaker is placed in the middle of two social relationships which have to be skilfully managed. For example, if the call has interrupted a face-to-face encounter (and we know that the telephone has long had a curious *de facto* priority in this respect, in the office context for example), the speaker may feel obliged to "play out collusive gestures of impatience, derogation, and exasperation" (*ibid.*: 221) for the benefit of the co-present other. Such gestures thus can be read as a kind of marker of inter-actional obligation. The mobile creates the situation in which, to an unprecedented degree, the likelihood is that the co-present other is a stranger to whom one's oblig-ations are more uncertain. The study of gaze and gesture can tell us about how in practice normative obligations are constructed, destroyed or more generally managed in practice, and can raise questions about the level and scope of per-ceived mutual involvement of speakers and involuntary listeners.

If the points made so far are principally focused on the possible intrusion of the private into certain kinds of public space, then turning our attention away from co-present interaction (as, perhaps, does the mobile user) highlights different dimen-sions. The mobile can also be a resource for personalising one's existence in public spaces, a resource for achieving privacy. It can for example, as one female teenage respondent has stated (and gender is obviously crucial for a more fully elaborated understanding of what the generalisations public and private might mean), provide a way of avoiding "strange encounters" in public spaces. It is not the only device for doing this, but it is a particularly effective one which visibly and audibly displays one's engagement with a remote other to those within earshot.[8]

As an aside: from the perspective of those within earshot, one grossly observable feature of mobile use in public spaces is that it has now become routine to hear just one side of a two-party conversation. Rippon and Ward's (2000) recent entertaining collection of "true mobile-phone conversations", is in fact not quite that but a col-lection of overheard contributions by one party. I wonder if there has ever before been published a book which reports one half of a number of conversations?[9]

To look at the phenomenon in these terms is to presume certain traditional (humanist) conceptions, notably of community and stranger, that some see as problematic and which certainly merit further interrogation. One might suggest, following Derrida, that there is a conflation of community and presence which needs deconstructing. Maffesoli (1996) argues that contemporary life is organised not around a rooted sense of community, but around multiple, loose and shifting forms of association and affiliation which he calls tribes. At an empirical level, we might ask whether there is such an affiliation between mobile users: whether, for example, the experience of having undergone embarrassment at receiving a call in an awkward situation creates a degree of empathy for others in a similar situation; whether, in terms of consumption and identity, there are forms of brand awareness with style connotations which lead to mutual identification; or to what extent the practice of short text messaging, so enthusiastically embraced within younger groups of users, itself constitutes a significant marker of social identity? At the conceptual level however, we have to ask how far the framework of public and private retains validity if we depart from humanist conceptions: for example, as we find in Simmel, of the private as interior subjectivity, and of the public as co-present collectivity?

This is of course not the only arena in which the public/private distinction has come under critical scrutiny. Weintraub (1997) and Bailey (2000) have noted the enormous semantic flexibility of the terms within social science; within philosophy, the distinction forms both the cornerstone and, for many, the fundamental flaw of Rorty's influential version of pragmatism (Rorty, 1989; Mouffe, 1996); while cultural critic Baudrillard argues, with typical hyperbole, that both public and private are being reduced and problematised by certain cultural developments, such as the fact that the most intimate features of personal life can now serve as "the potential grazing ground of the media" (Baudrillard, 1988: 20–21). More specifically, the notion of the private as interior subjectivity might be rethought as an effect of particular forms and norms of communication which, in turn, are derived from material assemblages of technologies, bodies and spaces (cf. Deleuze and Parnet, 1987).

The distinction emerges as a practical analytic problem when we observe the use of the mobile in many settings: for although public and private provide one way of beginning to think of the social significance of the phenomenon, there are many situations where, for example, it is not clear whether the use of the device should be taken to represent the intrusion of public into private, or of private into public.[10]

For these reasons, I suggest the mobile might be better thought of, in more general terms, as an indiscrete technology. This is not, primarily, because it facilitates forms of social indiscretion, although that is one view that can be taken, but rather because it has the capacity to blur distinctions between ostensibly discrete domains and categories, or more precisely to take its place among a number of social and technical developments that have this capacity: not only public and private, but remote and distant, work and leisure, to name but a few. It is not the case that these categories were unproblematic or given prior to the mobile; but the mobile provides one way of linking, and one route to rethinking them.[11] Nor, of course is it the case that distinctions are simply erased: but, where they are important, they may need to be built in different ways. Thus, whilst the mobile may permit undifferentiated access to work or, in cases where this is different, home, the mobile user can develop procedures for protecting that distinction. A proper understanding of this kind of "indiscretion" would therefore require empirical study of situated practical action, even if theoretical reflection can open up certain questions to be addressed. The need for this kind of balance is particularly true when it comes to considering one of this technology's apparently central affordances, mobility.

2.4 Mobility and Location

Mobility can be conceptualised in different ways, even though they all denote some form of movement in space and time. A distinction is made, by some mobile operators for instance, between three kinds of mobility which find their expression in the mobile phone/device: mobility of the user; mobility of the device; and, since they can be accessed from any point, mobility of services. This members' category seems a helpful way of thinking about mobility, in this context; but let us briefly look at some of the formulations of mobility that social theory has to offer.

Much of this takes the form of a discussion of the role of technology in relation to the changing nature and perception of space and time. Some theorists, such as Virilio (2000) see a shift from movement in physical space to relative stasis in which, thanks to electronic mediation "everything arrives without any need to depart" (*ibid.*: 20). It is possible to conceptualise a separate domain through which we move. The mobile, in its emerging form, would then offer the possibility of movement in both of these physical and virtual domains. There are of course good reasons to question this quasi-spatial metaphor for the use of electronic information, leaving aside the in some ways indeterminate current state of the mobile. The constraints of practical action will have a bearing on such a notion of simultaneous movement: in many settings for example, pedestrians can be observed coming to a standstill in this world in order to use some of the functions on their mobile.

The notion that the prevailing technology of a particular era can play a key role in modifying its sense of time is a well-established theme in social thought; Anderson (1983) for example documents the relationship between print technology, conceptions of nationality, and the emergence of a "modern" (secular) conception of temporality as linear time (and progress) in place of time as decay, as fall from grace. Many argue that we are in the process of another such transformation, begun in the early twentieth century but undergoing a further accentuation and radicalisation: what Nowotny in 1994 called "the illusion of simultaneity" (Roberts, 1998: 120) has replaced linear time. This means the replacement of one time by a series of overlapping times, and a corresponding intensification of the demands made upon people. The mobile would be a classic example of a device which both facilitates the demand and makes it possible to meet it.

Virilio however sees the growing availability of information on demand as leading to, not a frenzy of simultaneity, but a form of inertia: we will change from the "unbridled nomadism" of modernity to "the definitive inertia or sedentariness of whole societies" (Virilio, 2000: 20). The instant availability of all kinds of information at any time or place means that there will be no need for physical motion. If we are to accept this as a plausible vision, and there are very good reasons why we should not, the 3GM mobile is placed in an interesting position: at the juncture of virtual and physical mobility, it makes use of the electronic resources which would make the latter unnecessary. The staging of the workshop on which this book is based, in which all or most of the papers were submitted in advance for online availability, stands as a suggestive counter-example to this kind of implication that travel is unnecessary, or that the "event" (specific in space and time) is under threat. People have other reasons for continuing to travel, dwell or gather which merit consideration.

Virilio argues that alongside this sedentary tendency there is a process of being disconnected from our physical environment. The availability and increasing priority of remote communication and information is effecting a kind of spatial detachment from our immediate surroundings: "closer to what is far away than to what is just beside us, we are becoming progressively detached from ourselves" (*ibid.*: 83).[12] Again, as the previous section indicates, empirical study is needed to put some limits on such claims; and in any case his argument (1990 originally) predates the widespread use of the mobile, and the anticipation of its convergence with other technologies.

For example, notwithstanding claims that the availability of information from any point will make one's location irrelevant, one of the most grossly observable features of mobile conversations is precisely the degree of "situation-work" that is done. The utterance "Hello, I'm on the train", which has become a laughable cultural cliché in the UK, attests to this and to the fact that the socially competent mobile user needs to attend, often explicitly, to the contextual sensitivity of talk: information on whereabouts often serves to establish the grounds for the conversation in terms of constraints on and sensitivities with regard to possible topic, privacy, duration and so forth.

Convergence itself has, we are told, implications for our sense of space. One of the most cited uses for an enhanced device is the provision of location-specific information, or "location-based services" to use the current term. We may be in an unfamiliar location, but will be able to call a taxi without giving directions, or locate a nearby Italian restaurant (to use the kind of examples that are often repeated and which make use of particular assumptions about social milieus which could usefully be questioned). Virilio, and indeed McLuhan before him, have commented on the tendency to previsit locations, through one medium or another; to actually arrive somewhere is no longer surprising in the way that it was; and indeed Virilio suggests it is becoming replaced by prevision (*ibid.*: 22). Thus, according to this logic, the mobile would be one more technique by which the world became unsurprising;[13] but again this requires empirical qualification. The non-equivalence of location and situation may also turn out to be consequential for the developing form of these services: spatial coordinates being no direct guide to situational proprieties and requirements.

Finally, I want to comment on another aspect of the way in which the mobile might be said to affect our experience of mobility. The fact that the mobile allows people to be reached anywhere might contribute to a kind of stasis of identity for practical purposes; retention of the same mobile number means that moving office or house, being stationary or on the move is not significant from the perspective of the caller. The mobile user might be said to be perpetually on call. Nowotny argues that in a culture of simultaneity, the challenge for modern citizens is to "find time for themselves". The role of the mobile in this is uncertain, or rather not determinate. At a superficial level it appears to exacerbate the problem. More careful study, such as by Laurier (Chapter 4), suggests that using the technology to its full potential can, precisely, help to control time. Nevertheless, the perception that the mobile's potential is for continual availability is widespread, and thus merits some attention.

2.5 Being Available

One widely noted feature of the mobile is that it affords the possibility of perpetual contact.[14] This can be rhetorically configured in different ways, in particular as opportunity or demand, and in some cases in a thoroughly ambivalent manner that draws on both figures.

On the one hand being available can, in principle, be seen in a fundamentally positive light as part of the extension of transparency described above. Representations of the mobile and of the behaviour it facilitates, in advertising for instance, tend to draw on this sense of communication as, in itself, empowering and solving problems caused by being out of touch. As such they form part of the vision of a "transparent society" as ably analysed by Vattimo (1992). In such a vision, increases of communication are given a normative inflection, and articulated with notions of community; this represents, argues Vattimo, a romanticised ideal in which community is achieved by unrestricted communication (see also Moran-Ellis and Cooper, 2000).

On the other hand, uninterrupted availability can be described as a potentially oppressive feature of the social effects of technology:

> This "machine", operated in the closest vicinity to the word, is in use; it imposes its own use. Even if we do not actually operate this machine, it demands that we regard it if only to renounce and avoid it. This situation is constantly repeated everywhere, in all relations of modern man to technology. Heidegger (cited in Kittler, 1999: 200)

Heidegger's comments seem apt for certain current discussions of the imperative to get connected. Despite the widespread uptake of the mobile in recent years in the UK, this concern continues to be expressed from time to time. Newspaper articles discuss it. For example The Guardian (5 April, 2000) ran an article by columnist Julie Burchill which was trailed on the front page of the section under the heading "Why I have never bought a mobile phone". The fact that this was considered newsworthy itself says something about the current level of expectations surrounding the mobile, and the fact that we are forced "to regard it if only to renounce and avoid it". The layout of the article, with quotations from non-users accounting for their non-use reinforced this impression.[15] Certain advertising plays with it: a 1999 Carphone Warehouse radio advert aimed at the mobile professional, for example, concluded by acknowledging that perpetual contact can be bad news for some with the words "You can run but you can't hide". Availability here is acknowledged as both an advantage and a possible liability. At the same time, when we take into account that the "machine" that Heidegger is so vehemently castigating for its corruption of Being is the typewriter, we may think a degree of caution is in order. With hindsight, resistance to technology can often look rather quaint.[16]

There is more than one strand to follow here however. As noted earlier, the fixed line phone already assumes a form of priority over face-to-face interaction in many settings: the phone may be ignored, with difficulty, but it cannot be kept waiting, unlike the co-present other who may be mollified with sympathetic gestures during the call. The mobile imperative takes different forms. First, one may be called to account for non-ownership. Secondly, owners may be called to account for non-use, having them switched off, or for whatever reason being unavailable. (These forms of accountability may themselves be related to Nowotny's notion of the demands of simultaneity.)

The terms in which anxieties are expressed about this second strand recall Heidegger's view of modern technology as something which "enframes"; loosely, as something which converts the world into a resource to be utilised. His well-known example is of a hydro-electric plant which changes the status of the river Rhine into something purely instrumental, something which has utility (Heidegger, 1977). The analysis can equally be applied to the conversion of people to resources. Ronell's typographically extraordinary *The Telephone Book* (1989) pursues this kind of critical line, suggesting that to answer the phone is to be part of an insidious and instrumental technologising of beings.[17]

This "demand" can however be seen in slightly different terms. Derrida's analysis of aspects of Joyce's *Ulysses* comments on the use of the telephone in that text, and we can cautiously infer, more generally:

> Before the act or the word, the telephone. In the beginning was the telephone [...]
> There are several modalities of the telephonic yes, but one of them, without saying
> anything else, amounts to marking, simply, that we are *here*, present, listening, on
> the end of the line, ready to respond but not for the moment responding with any-
> thing other than the preparation to respond. Derrida (1992: 170)

Here, by a kind of supplementary logic, the telephone is not a technological (and problematic) addition to communication but something presupposed by communication; and, as such, can then be(come) a signifier of communication more generally. (Again this somewhat abstract idea finds an analogue in certain advertising strategies: one thinks of the ways in which British Telecom in recent years have begun to position themselves as proprietors of communication per se, from the "Good to Talk" campaign to their "Talk Zone" at the Millennium Dome.) Thus the telephone, to a greater extent, represents not so much technological rationality but, in a particularly graphic form, sociality *per se*.

It is important to note, by way of qualification, that even Heidegger's formulation should not be read as suggesting the determinate effects of particular technologies, for his concept of the technological is a broad one which goes beyond specific technologies, and is seen as the basis of our modern instrumental world view: the typewriter, the mobile or whatever are manifestations of this. The point of interest here, and elsewhere in this chapter, is in the assemblages of technology and normative social practice that both grow around and shape them. Following Serres, we might describe the mobile as a "quasi-object" (Serres, 1995) which is neither constitutive of, nor reducible to the ways in which it is used. It is interesting in this respect to consider forms of perpetual contact which predate the mobile. For example, in Woody Allen's early film *Play It Again Sam*, one character, a businessman, spends most of the film ringing in to his secretary to let her know which number he can be reached on, on one occasion telling her the number of a call-box he will be passing in between appointments: yet the comic effect of this indicates and results from the absurdity of this form of practice given the prevailing technology available at the time.

2.6 Conclusion: Mobile Theory

I have tried to bring some ideas from social theory into contact with what is, in some senses, a mobile and transitory technology in order to see what, if anything, this can tell us about what I still consider to be a remarkable social phenomenon. I have tried to indicate in passing the limitations of some of this kind of theorising, and the empirical work that informs other chapters in the book will make some of these limitations even more manifest. If the mobile (or the mobile conversation) is an elusive and disposable phenomenon, then theory which attempts to explicate it might be treated in the same way. Deleuze states that "concepts are exactly like sounds, colours or images, they are intensities which suit you or not, which are acceptable or aren't acceptable" (Deleuze and Parnet, 1987: 4). Without endorsing this as a general model, it may be an appropriate description of the motivations behind this chapter.

However, I also suggest that the mobile is strategically useful in that it has the capacity to raise questions about the appropriate form and level of analysis. It is a technology which connects the global, in the form of a network of satellites and transmission points, with the most local of social interactions; in other words, it forms a juncture, a point of contact between different domains, each of which lend themselves to a different kind of analysis. It may therefore be said to be indiscrete at the theoretical level, and a kind of theoretical flexibility or mobility may be required in order to do it justice. We can choose to look, with Goffman or the ethnomethodologists, at modes of conduct in everyday life and the norms and obligations to which they are subject, keeping our gaze primarily on local interaction. Following Latour,[18] we may take the unit of analysis to be the hybrid entity or, to use Deleuze's term, the assemblage of person and technology. Here, in an explicitly post-humanist analysis, the mobile actor becomes the point of intersection of diverse associations and networks which are not localised but exist to enable "action at a distance".[19] Alternatively, certain categories of phenomena may suggest that a systemic level of analysis is appropriate. For example, the mobile may be thought to facilitate communication and a degree of reflexivity through a body or group of people: witness the apparently key functional role of the mobile in organising recent European industrial action over fuel costs, often with no recourse to existing structures of organised labour.

Convergence may have further implications for ascertaining the appropriate mode and unit of analysis:[20] for not only does it imply the bringing together of certain technical functions, but it also brings together different bodies of literature which may not coalesce in any simple sense. So, within science and technology studies, the opposition to technological determinism is clearly articulated; yet, in this domain, this stands in an antagonistic relationship with media theory's insistence that particular media are never simply neutral carriers of semantic content.[21] This tension is to be welcomed: it provides an opportunity, in some ways creates the necessity for theoretical refinement. The extent of theoretical mobility in this area may provide one useful if indirect measure for assessing the emerging social significance of the mobile.

Notes

1. The project, "The socio-technical shaping of mobile multimedia personal communications (STEMPEC)", is funded by the ESRC (L487 25 4002) and Vodafone, One2One, Orange, BTCellnet, and Granada Media Group, as part of the Foresight Link Programme. I am grateful to Nicola Green and Ged Murtagh for their fieldwork and, together with Richard Harper, their discussion of some of the issues that are touched on here.
2. Castells (1996) and Urry (2000) provide good examples of sociological literature which addresses highly pertinent issues but stays at a certain level of generality which we might call macro. For a discussion of the problems of seeing this distinction in purely analytic terms, and of the value of focusing on the substantive movement between these levels, see Callon and Latour (1981). I return to this issue later in the chapter.
3. See for example Bijker *et al.* (1987) and Grint and Woolgar (1997).
4. The uptake of mobiles in Europe is far higher than in the USA.
5. Some have argued that other media, such as television, have been significant for changing the boundaries between public and private: see Meyrowitz (1985).
6. For exploration of other aspects of the relation between urban life and the mobile, see Chapter 5.
7. This tension may also take the less subtle form of explicitly negative reactions as documented by Ling (1997) and Haddon (1998).
8. The audible and the visual are two properties of the mobile which may form part of different configurations of behaviour.
9. But cf. Cocteau's (1992, originally 1930) dramatic exploration of one side of a telephone conversation. Thanks to Elizabeth Churchill for this point.
10. Thanks to Sally Wyatt for clarification of this difficulty.
11. In other words, the mobile may also be indiscrete with respect to alternative forms of analysis.
12. This is an interesting re-inflection of Benjamin's notion of "aura" in which certain things which are immediately present (such as works of art) retain a kind of remoteness, mystery and distance: here the remote becomes present, and the present becomes not mysterious but secondary or a matter of relative indifference.
13. More prosaically, we might note the way in which towns are themselves becoming more uniform and unsurprising.
14. See for example contributions to the conference, "Perpetual Contact", Rutgers University, New Jersey, 12–13 December, 1999.
15. The article ran under the headline "Slaves to the Mobile".
16. One quails at the thought of Heidegger having to deal with an animated paper clip interrupting him to guess, incorrectly, that he is writing a letter.
17. The potential monitoring function of the mobile is clearly pertinent to this line of argument: see Chapter 3.
18. See, for example, Latour (1994).
19. As indicated above (note 2) conceptualising action in these terms is not the same as looking at, for example, information flows at a purely, and relatively unmediated, macro-level (Castells, 1996).
20. To return to Latour's "immutable mobile", convergence would in principle give the mobile actor access to all the information on the web from any location: this very lack of inherent selection might be said to distinguish this technology from his concept of a specific inscription or representation which allows action at a distance; or it might represent an ultimate form of it, albeit one in which the concept's very generality diminishes its analytic value. On the problems of total recall and the need for selection, see Umberto Eco's contribution to Eco *et al.* (1999).
21. Some media theorists however argue that the convergence resulting from digitalisation represents a qualitative break, and will "erase the difference between individual media" (Kittler, 1997: 31–32).

References

Adorno T (1973) *Negative dialectics*. London: Routledge.

Anderson B (1983) *Imagined communities*. London: Verso.

Bailey J (2000) Some meanings of "the private" in sociological thought. *Sociology*, 34 (3), pp. 381–410.

Barthes R (1993) *Camera Lucida: reflections on photography*. London: Vintage.

Baudrillard J (1988) *The ecstasy of communication*. New York: Semiotext(e).

Benjamin W (1903) *Charles Baudelaire*. London: Verso.

Bijker W, Hughes T and Pinch T (1987) *The social construction of technological systems*. Cambridge, MA: MIT Press.

Callon M and Latour B (1981) Unscrewing the big Leviathan, or how do actors macrostructure reality? In Knorr-Cetina K and Cicourel A (eds) *Advances in social theory: toward an integration of micro- and macro-sociologies*. London: Routledge.

Castells M (1996) *The rise of the network society*. Oxford: Blackwell.

Cocteau J (1992/1930) *The human voice*. London: Samuel French.

Deleuze G and Parnet C (1987) *Dialogues*. London: Athlone.

Derrida J (1992) Ulysses Gramophone. In *Acts of literature*, New York: Routledge.

Eco U, Gould S, Carriere J-C and Delumeau J (1999) *Conversations about the End of Time*. Harmondsworth: Allen Lane.

Frisby D (1985) *Fragments of modernity*. Cambridge: Polity.

Goffman E (1971) *Relations in public*. Harmondsworth: Allen Lane.

Grint K and Woolgar S (1997) *The machine at work*. Cambridge: Polity.

Habermas J (1989) *The structural transformation of the public sphere*. Cambridge: Polity.

Haddon L (1998) Il Controllo della Comunicazione. in L Fortunati (ed.) *Telecomuncando in Europa*. Milano: Franco Angeli.

Heidegger M (1977) *The question concerning technology and other essays*. New York: Harper & Row.

Kittler F (1997) *Literature, media, information systems*. Netherlands: G+B Arts.

Kittler F (1999) *Gramophone, film, typewriter*. Stanford, CT: Stanford University Press.

Latour B (1986) Visualization and cognition: thinking with eyes and hands. *Knowledge and Society*, 6, pp. 1–40.

Latour B (1994) On technical mediation. *Common Knowledge*, 3 (2), pp. 29–64.

Ling R (1997) One can talk about mobile manners! The use of mobile telephones in inappropriate situations. In Haddon L (ed.) *Communications on the move: the experience of mobile telephony in the 1990s*, COST 248 Report. Farsta: Telia.

Maffesoli M (1996) *The time of the tribes*. London: Sage.

Marvin C (1988) *When old technologies were new*. Oxford: Oxford University Press.

Meyrowitz J (1985) *No sense of place: the impact of electronic media on social behaviour*. Oxford: Oxford University Press.

Moran-Ellis J and Cooper G (2000) Making connections: children, technology and the national grid for learning. *Sociological Research Online*, 5 (3).

Mouffe C (ed) (1996) *Deconstruction and pragmatism*. London: Routledge.

Rippon A and Ward A (2000) *I'm on me mobile*. London: Robson Books.

Roberts R (1998) Time, virtuality and the goddess. In Lash S, Quick A and Roberts R (eds) *Time and value*. Oxford: Blackwell.

Ronell A (1989) *The telephone book*. Lincoln, NB: University of Nebraska Press.

Rorty R (1989) *Contingency, irony and solidarity*. Cambridge: Cambridge University Press.

Sacks H (1992) *Lectures on conversation*, Vol 2. Oxford: Blackwell.

Schegloff M (1992) Introduction. In Sacks H (ed.), *Lectures on conversation*, Vol. 1. Oxford: Blackwell.

Serres M (1995) *Genesis*. Ann Arbor, MI: University of Michigan Press.

Urry J (2000) Mobile sociology. *British Journal of Sociology*, 51(1), pp. 185–203.

Vattimo G (1992) *The transparent society*. Cambridge: Polity.

Virilio P (2000) *Polar inertia*. London: Sage.

Weintraub J (1997) The theory and politics of the public/private distinction. In Weintraub J and Kumar K (eds) *Public and private in thought and practice*. Chicago, IL: University of Chicago Press.

Chapter 3

Who's Watching Whom? Monitoring and Accountability in Mobile Relations

Nicola Green, Digital World Research Centre, University of Surrey, England

---

3.1 Introduction

This chapter offers some tentative thoughts on an issue that has recently become a subject of public debate – the capability of mobile technologies, especially emerging location-based services, to act as technologies of "surveillance". I take as my starting point two instances of social relations, one drawn from what might be termed "popular culture", the other drawn from observational research.[1]

The first of these instances is an advertisement for a mobile phone network which portrays a group of friends watching a young woman in a variety of locations as she goes about her everyday life, and communicating her activities to each other via their mobile phones. The activities of "watching" and "following" are supported in this advertisement by the use of mobile technologies to monitor the tastes and preferences of this young woman, and to organise "accidental" meetings with one of the young men. This "surveillance" begs the question – who is now watching whom, and through what technologies?

The second instance is a ubiquitous feature of talk one encounters when observing the use of mobile phones – the question "where are you?" This short phrase, so simple as to seem completely unremarkable, is so widely used that its presence deserves attention. In his recent paper "Why people say where they are during phone calls", Eric Laurier (1999) suggests that such questions of geographical location establish mutual contexts for communication, and enable shared circumstances between the parties who are communicating at a distance. I'd like to suggest in the course of this chapter that this phrase also serves to establish relationships of "mutual accountability" and trust through gathering information about the physical, social and psychological conditions of those with whom one is relating. Moreover, this accountability is a feature both of the relationship established via the mobile technology, and also a feature of co-present social relations when the technologies are used in public space. Monitoring of location and activities in this instance serves both to cement personal or intimate relationships and to make an individual's activities transparent, visible and accountable to both co-present and tele-present others.

What is interesting about these examples is that the practices of information gathering about others, and assumptions about who does it and to whom, seem to have undergone a shift through the uptake of emerging mobile technologies.

Information gathering activities seem now more widely "normalised", and more often taken for granted as resources in everyday relations of trust. These emerging social relations require a change in the conceptual apparatus through which information-gathering activities are currently documented and understood by social scientists. One of the most often used means of theorising information gathering in modernity, the notion of "surveillance", seems unable to account for new webs of connection and availability, and the expectation of transparency of activity, and visibility to others, in everyday life.

"Surveillance", a term most often associated with the law enforcement apparatus of the state, can be used to theorise some important new changes in mobile information gathering and communicative availability, most particularly in the relationships been individuals and corporate social bodies such as the state or corporate businesses. It is more difficult to employ this term, however, with regard to the information gathering and communicative activities now routinely carried out amongst individuals, in new forms of social relation previously unavailable to the same degree. Rather than the state and institutions surveilling populations, populations are also "surveilling" themselves and each other through new, mobile technologies in the course of intimate and interpersonal everyday relations. This normalises the notion that individuals *should* be available and accountable to others, visibly and transparently, at any time and place (Green and Harvey, 1999).

The aims of this chapter are therefore twofold. The first is to trace new relations between corporates and individuals, and between individuals in interpersonal relationships, in information gathering activities that are newly presented by emerging mobile technologies. The second aim is to explore the uses and limits of the concept of "surveillance" to describe these activities, and to suggest alternative conceptual frameworks, integrating notions of mutual monitoring and accountability, that account for these new, increasingly complex relationships.

Before engaging with some of the research material of relevance, a brief foray into what is meant by "surveillance", and an exploration of the differences between this term as a conceptual framework and its use to describe social practices, would be useful. I am approaching this discussion with the assumptions that technologies of surveillance are not in themselves "good" or " bad", but neither are they neutral, and that what is required is a nuanced understanding both of the surveillance *capacities* of new digital technologies (Rule, 1973) and of the ways they are socially shaped in everyday practice.

3.2 Surveillance and Mobile Technologies

The notion of surveillance has become commonplace in social and political thought in the past 20 years to describe the power relationship between the state and the individual, most often with reference to the work of Michel Foucault. In *Discipline and Punish* (1979), Foucault indicated the significance of surveillance for the monitoring and control of modern populations. This work, which investigated how the exercise of state power has changed over the period of modernity, took as its starting point the evolution of the modern prison system.

Foucault's argument stated that whereas the governance of populations was historically enacted through physical force and centralised hierarchies of power, populations are now governed by the state through techniques of "discipline". These regulatory modes of social behaviour require both the consent of the governed, and their *self*-regulation via various technologies – not least of which are technologies of information and communication.

The relations of surveillance are a central mechanism generating this disciplinary self-regulation amongst populations. Foucault (1979) used the image of Bentham's panopticon prison design to illustrate how surveillance – or even the possibility of it – creates self-regulating populations. Bentham's design for a prison placed a guard tower in a central space, with the prisoners confined in cells arranged in a circle around the tower. With this design, the guards can see the activities of any and all prisoners, at all times. The actions of the prisoners are rendered utterly transparent and visible to the guards, and the consciousness the prisoners have of being watched transforms their actions. In time, guards are no longer needed, as the prisoners have internalised the watchfulness of the guards and become self-regulating. According to Foucault (1977: 155):

> there is no need for arms, physical violence ... Just a gaze. An inspecting gaze, a
> gaze which each individual under its weight will end by interiorising to the point
> that he [*sic*.] is his own overseer, each individual thus exercising this surveillance
> over, and against himself.

Extending this model of the prison, Foucault argued that contemporary states use technologies of visibility, primarily in the form of "information", to achieve surveillance of their populations, and use the principle of *uncertainty* to exert micro-level relations of power in everyday life (Lyon, 1994). This thesis furthermore fundamentally implicated the production of *knowledge* in the power relationship. Catalogues of individual and personal details are kept in extensive centralised databases which record everything from a name, address and bureaucratic numbers, to the records of individual behaviour (such as criminal records (Foucault, 1979)), or the records of an individual body's functioning (such as medical records (Foucault, 1973)). The effect of these surveillance practices is not only the effective regulation of populations, but also the institution of state regulatory practice as social norm. When the recording of information in this way becomes a normally accepted practice to which individuals consent,[2] social subjects become *self*-disciplining to the norms thereby inscribed. Here, the concept of surveillance intersects with those of "discipline" and "normalisation", and more recent theorists have developed Foucault's theories of power through visibility to argue that micro-social practices between individuals grounded in desire and identity are as important as state catalogues when it comes to individual self-regulation of behaviours. According to Sawicki (1991: 67–68):

> [d]isciplinary practices ... are located within institutions ... but also at the
> microlevel of society in the everyday activities and habits of individuals. They
> secure their hold not through the threat of violence or force, but rather by creat-
> ing desires, attaching individuals to specific identities, and establishing norms
> against which individuals and their behaviours and bodies are judged and against
> which they police themselves.

The desire for *identity* with others incites the desire to maintain self-surveillance, and self-correction to social norms (see, for example, Bordo, 1989, 1993; Lyon, 1994; Sawicki, 1991). According to these theorists, the dynamics of social power reside as much in micro-social interactions and interpersonal relations as they do in the mechanisms of institutions or corporate bodies, and the power generated through information and communication technologies resides between individuals, and between individuals and institutions, in the use of technologies in everyday life. Complex dynamics of the exercise of power and resistance to it are thereby generated.

We might easily think of contemporary information and communication equivalents to Bentham's panopticon beyond the prison, in wider social life (Bogard, 1996; Loader, 1997). Besides the now ubiquitous surveys of populations routinely carried out by both state and private enterprises, new media and information technologies offer numerous instances of centralised surveillance techniques. Technologies such as CCTV (closed-circuit television) are a case in point. CCTV is centralised, it exerts an all-encompassing gaze, it is linked via telecommunications networks and cross referenced amongst a number of databases (Lyon, 1994), and its intended effects are to generate self-regulating individuals. The proponents of CCTV argue that crime rates in areas covered by CCTV have gone down,[3] and its use in public space has become largely taken for granted, at least in the UK (McGrail, 1997; McCahill, 1996).

The existence and use of CCTV in policing practices conforms very closely to Bentham's view of the panopticon, but Foucault's use of Bentham is only part of a wider theory of micro-level power relations that rely not only on the production of surveillance technologies and use of them by states in carceral punishment, but also the reception and collusion with technologies of knowledge and visibility in everyday life. Indeed, CCTV is only one amongst an increasing range of technologies that hold the potential to intensify practices of surveillance. Contemporary theorists therefore argue that along a number of measures (beyond the model of the panopticon), surveillance is intensifying along with the convergence of information and telecommunications systems which require a rethinking of the concept of surveillance itself (Lyon, 1994). Those measures of surveillance capacity include the storage capacity of computers, the comprehensivity of their reach, the speed of the flow of information, and the relative visibility of persons in the network (Lyon, 1994: 52). Ironically, the "new surveillance" is said to acquire more routinsed and comprehensive power through its distribution and decentralisation, than through the centralisation of databases (more highly networked and distributed systems achieving greater social penetration of monitoring). According to some, the convergence capacities of these technologies can produce a "new surveillance", which:

> transcends distance, darkness and physical barriers ... transcends time ... is of low visibility or invisible; data subjects are decreasingly aware of it ... is frequently involuntary ... prevention is a major concern ... it is capital-rather than labour-intensive ... [and] involves decentralized self-policing. (Lyon, 1994: 53)

Equally, such a "new surveillance" involves consumption as well as production. What convergence mobile technologies highlight, situated as they are amongst a range of technical innovations such as the internet, email and the world wide web, is the extent to which state surveillance has now been complemented by information gathering activities on the part of commercial institutions. Here, we might think about the ways that commercial institutions gather information about their customers, compiling extensive databases of information about consumer preferences and habits, as well as exhaustive lists of personal details. These practices have also become taken for granted, and often employ the active collaboration of those under scrutiny to maintain and update the information in a form of self-surveillance (Lyon, 1994).

Dale Spender (1995), among others, argues that those very technical systems which allow people to *communicate*, also allow them to be *monitored*, which indicates a recursive process of ambiguous social shaping of information and communication technologies in both their production and consumption. The "cookies" that have become standard and almost ubiquitous in websites, for example, do allow consumers to explore a wider range of information resources of the kind they want access to. At the same time, the cookies also provide information to host servers about consumers' visits to those sites, and the click through rates on pages are routinely monitored to provide information not only about the number of "hits", but navigation histories, time spent in particular site areas, and even personal details of purchases made through those visits. Email can be monitored by all kinds of public and private agencies (both the state and employers) through records of central servers. Similarly, the records of current mobile phone use provide a detailed temporal and geographical history of use, the frequency and duration of use, and the numbers called and received, all correlated to particular devices. In certain (rather exceptional) circumstances, this can be used by state authorities to establish the location and use of particular devices on the assumption that those devices are attached in some way to particular people. Records of phone calls can be investigated in cases of harassment, for example, on the request of the police under the Data Protection Act 1998 in the UK.

These examples, and the conceptual apparatus that addresses them, point to another useful distinction to keep in mind when discussing the potential surveillance capacities of particular technologies. This is Giddens' distinction between the bodily surveillance of individuals (as in CCTV) and the visibility of individuals as represented by their "digital personae" in database records. Both of these devices may be employed by the state – the latter is likely to be more often employed by private enterprise. These different forms of surveillance become important when considering the cluster of people, technologies and organisations that comprise increasingly convergent mobile technologies.

In contemporary mobile technologies, networks of individuals and communities are connected via a technological infrastructure, increasingly personalised devices (possessed by and associated with particular individuals by all parties), and a range of organisations that provide devices and services. With the introduction of location-based and satellite services, the technical capacity for surveillance holds a far more comprehensive reach in framing not only the relationships between organisations and the individual, *but also between individuals* – a set of relationships relatively neglected in research specifically focused on relations of monitoring. One of the most often cited uses of new location-based services is the ability to locate the geographical position of friends and loved ones while on the move in large urban spaces.

On the one hand then, mobile devices do present the possibility of (public or private) institutions actively monitoring the activities of individuals, and relating to them on that basis in similar ways to other contemporary digital technologies. While giving mobile phones to job seekers, for example, may mean that they have further information resources with which to seek work, it is equally possible that possession of those devices might be used to hold individuals institutionally accountable for their day-to-day activities, with the institutions monitoring their responses to the information provided. Similarly, the convergence of information and communication functions in the same devices offers similar opportunities. Geographical information provided to individuals through satellite services might provide a tool with which to carry out their everyday lives in more convenient and time-saving ways. At the same time, it is equally possible that those same devices might be used to monitor the geographical whereabouts of that individual in order to provide unsolicited advertising about burger bargains at the McDonald's they are walking past. In both these cases, the monitoring activities can come to be taken for granted, and become "normalised" as an aspect of social life to be accepted and integrated into everyday interaction. There are therefore ways in which these devices can be used to (either "publicly" or "privately") institutionally monitor individuals, and for individuals to resist that monitoring through their use of the very same devices.

While the information that can be gathered via the use of mobile devices can be deposited centrally in institutions, these technologies also allow for webs of communication – multiple connections between individuals and institutions, such as the example of the friends monitoring the young woman through their use of mobile phones. Like the internet, the physical form of these technologies means that they, and the opportunities for communication presented by them, are both distributed and unfixed. Moreover, the social contexts in which mobiles operate are constantly changing, in contrast to the relatively fixed and centralised relationships described by Foucault.

So far, despite the shift in social scientific conceptualisations of surveillance towards "consumption", few studies have examined the potential relationships between the predominantly abstracted and bureaucratised data-generated and bodily surveillance of individuals by organisations, and the tele-present monitoring of others that is based on intimate, face-to-face relationships of those who are known to others in communities, and are accountable to each other in relations of mutual trust and accountability. To explore further the possibilities in this train of investigation, I want to turn to some of the empirical cases we have thus far encountered in our fieldwork to explore how surveillance and accountability might usefully be drawn together. My assumption throughout this argument is that qualitative, empirical investigations of "talk" about mobile technologies and the relationships mediated through them indicate that "surveillance", as a practice rather than a concept, is contextually dependent, and highly reliant on changing notions of the relative intimacy or abstraction of the relationship concerned.

3.3 Who's Watching Whom?

With these themes in mind, I'll explore some research materials gathered in interviews with teenagers, as well as observations of use in public spaces. My argument is that teenagers are visibly subject to both institutional surveillance and parental and peer monitoring, and so make an interesting case study in the exploration of the relation between surveillance and accountability. They are subject to this monitoring (and participate in it themselves) in ways that make the growing importance of mobile technologies in everyday life obvious. Mobile devices can both facilitate the regulation of individuals (not limited to teenagers) by institutions, and facilitate monitoring by significant others, but they are also devices that can be used to resist surveillance and monitoring by others. The dynamics of these relations seem to be changing alongside change in mobile technologies.

In the research work carried out thus far on the Stempec project,[4] teenagers have provided a rich source of data about social practices of mutual co-dependence, trust and accountability in everyday life, especially as these are carried out via mobile technologies. The category "teenager" is a somewhat flexible term in everyday use, because its use as an identity and behavioural category (as well as and distinct from its use as a descriptor of age) contains multiple meanings and is contextually based, and changes with time, place and social and interactional circumstance. Because "teenager" lies between the social categories of "child" and "adult" that frame it, their status with regard to independence of action is sometimes questioned by the "others" (e.g. parents, school institutions, the law) who are held socially responsible for their welfare.[5] Young people may therefore become subject to various forms of monitoring from both family and social institutions – a set of social relations which is both "normalised" and taken for granted. Increasingly, mobile phones are becoming embedded in these relations, both as devices used for parental monitoring, and objects that are institutionally monitored by schools as part of the surveillance and control of school-based youth behaviour.

There are two kinds of monitoring to which teenagers and their activities with mobiles are subject. The first is monitoring the teenager's state, whereabouts and activities through communicating via the mobile technology. Mobile phones are increasingly used by parents as devices to establish the condition and whereabouts of teenagers in possession of them, and maintain relations with their children when they are in places hitherto inaccessible. These relations are based on mutual trust and accountability between parents and their children. Teenagers' accounts detail the importance of communication with family members as one of the primary reasons to have a mobile phone, and cite parental (and their own) concerns over "emergencies" and "safety" as central reasons for carrying and using a phone. At the same time, teenagers sometimes acknowledge that they will allow themselves to be subject to potential monitoring by parents in cases where they feel it is undesirable, simply to gain possession of a mobile phone. They manage mutual accountability in such instances by not quite telling the whole truth about where they are and who they are with – or they will lie.

The second form of monitoring to which teenagers are subject is in their use of mobile phones, where the object itself and the activities carried out through it become the focus of parental and school/institutional monitoring activity. The teenagers' own accounts describe the domestic rules and regulations for mobile use, including when, where and for how long the mobile phone can be used in both domestic and public settings. According to the teenagers, parents monitor their mobile usage and the cost of it (and regulate use of the landline phone in relation to it). One of the most important reasons for using the mobile phone within the home is the privacy it affords teenagers. Talking to their friends and conducting their social affairs can all be done in the privacy of their own bedroom, because within the home the landline phone does not always provide for a private conversation. This indicates that teenagers use mobile communications to counter or avoid the parental monitoring of their affairs and the gate-keeping activities of parents on the landline phone in common domestic spaces.

Teens were not only called to be visible and accountable to their parents in mobile relations, but are also subject to (a more extensive) monitoring and regulation while at school. The teenagers describe institutional rules regarding mobile phone use in schools, which are banned from lessons and sometimes banned altogether. Whereas parents were concerned with a generalised "danger" presented by public places, both teenagers and their schools identified "danger" or "risk" in ownership of the devices themselves. The school ban on phones gives institutional recognition to the fact that younger children perceive risk in carrying phones, as they might be stolen from them to support a school-based black market for consumer goods. The bans on mobile devices fitted into the existing regulation of banned objects (such as leather jackets, trainers and Walkmans), as well as existing regulations of behaviour (they were not allowed to go down certain streets on the way home from school for reasons of safety, for instance). When mobile phones were found by teachers, the devices were confiscated and returned at a later date.

Unsurprisingly, the teenagers resisted this institutional surveillance of their activities, and continued to use mobile phones within schools. They contested the school's definition of danger, and used the definitions arising from parent's concerns to argue against the ban. The qualities of the mobile device itself become important means to resist monitoring by teachers. Devices which are most easily concealed from view have the most value for teenagers, and functions such as text messaging which can be carried out in silence are described as "discreet". Furthermore, the teenagers would use private spaces within schools to use their mobile phones. One strategy employed is to use the phones in the school toilets, as well as public places outside the school grounds on the way to and from school. One young girl described situations where, going into the girls' toilets, she would always find a group of girls talking on their phones – finding one of the only private places in the school away from the surveillance of teachers.

Monitoring and regulation by adults, both in the home and within institutions, is therefore both supported and resisted by the teenagers themselves in moves towards independence and control of their own affairs. Mobile devices thus provide a site of negotiation for monitoring, regulation and mutual accountability. Teenagers develop "parent management strategies" in response to parental surveillance of mobile use, in much the same way as they resist institutional monitor-

ing of mobile use by creating private spaces of mobile phone use away from surveillance by teachers.

The significance of these examples for the understanding of "monitoring activities" in everyday life is that it indicates that mutual accountability is linked to "surveillance" in the sense that what constitutes relations of trust, or relations of
regulation via monitoring, is *contextually* based. The experience and practice of
"surveillance", as describing social relations of monitoring and visibility for purposes of regulation, depends on pre-existing social relationships and the contexts
in which they are called into social action. Whereas some instances of teen-parent
relations might be understood as trust and mutual accountability, it can be *experienced* in some circumstances by teens as "surveillance" by those with relative
power in their lives. This adds credit to the notion that "surveillance" is not *one*
predetermined and institutionally based set of relationships that rely on specific
technologies, but a range of social practices and understandings mediated through
a range of technologies in everyday life, and closely bound up with relations of
mutual trust and accountability.

So how might this suggestion be explored further in relationships of mutual
monitoring and accountability in relations mediated through mobile technologies
with "surveillance" capacities? The case of teenagers makes obvious a more general
shift that adds a number of mutual monitoring and accountability relations to
more institutionalised forms of surveillance studied in the past. In the case of
adults, this involves a shift from the *active regulation of other persons* (whether in
the realm of the state or in consumption) via mobile networks, to the mutual
assumption of their *self-regulation* as persons accountable for their use of mobile
devices, and visible and available to significant others at all times and places.

Indeed, teenagers amongst themselves use exactly this kind of mutual and self-
regulatory accountability when using the devices to establish relationships of
status and community amongst themselves (rather than between themselves,
parents and schools). Quite apart from the use of the devices for voice or text communication, monitoring *other teenagers'* possession of mobile phones, the brand of
phone and service, and the phone's aesthetics and functionality-in-use (such as the
names contained in phone book/contact lists), are important ways that teenagers
negotiate their status positions within complex teenage community relations.
Furthermore, they use the devices to monitor complex and highly dynamic shifts
of peer relationships – where their peers are, what they are doing and, most importantly, who they are with (see Chapter 7).

Among adults, these kinds of monitoring and accountability – where others'
phones and their activities with them become the focus of monitoring – also exist.
In our research thus far, activities of use in public spaces provide rich examples of
mutual surveillance where individuals are potentially held accountable for their
use (or non-use) of the phone. Individuals' awareness of the possibility of public
censure of mobile phone use is apparent in the observed activities of individuals
to create boundaries between public and private interaction in public space, and
regulate their own behaviour – expressed in gaze, facial expression, or even in
speech (see Chapter 6). In our research on trains and in stations, when individuals
do use their phone, the gaze and gesture they use to create private space demon-

strates their awareness of co-present context, the normal ordering of public behaviour and the moral accountability of that behaviour. Changes in behaviour, such as leaving a seat on a train or changing seats, crouching the body over in a seat, or averting the gaze in "civil inattention", are all strategies employed by users to create "fictive boundaries", and demonstrate awareness of others' presence and potential monitoring (Ling, 1999).

The other way institutions and individuals engage in surveillance practices through mobile devices is in the relations of monitoring and accountability through the communicative functionalities of the devices themselves. As in the case of teenagers, adults and the institutions to which they are affiliated monitor others in their interpersonal and organisational groups by using the devices to communicate, and engaging in their own self-regulation by making themselves available for tele-present interaction. It has been demonstrated in social and legal research that some employers, for example, routinely monitor their employees' information and communication activities (Dichter and Burkhardt, 1996; Mason *et al.*, 1997) through the functionalities of the technological devices. An awareness on the part of employees of this institutional monitoring (or its potential) through mobile devices is demonstrated in research carried out on mobile teleworking. In a study by Hill *et al.* (1996), mobile teleworkers' perceptions of the impact of mobile devices on work and family life were examined. The study reported that some workers felt that the advantages of mobility and "telepresence" were sometimes offset by the drawbacks of permanent availability, institutional visibility and accountability in work. The constant presence of mobile work technology in the domestic setting, and the potential regulation of activities and obligations it represented, impinged on home and family life. One implication of this finding is that the presence of the employer becomes embodied in the devices, and that the presence of such devices prompts self-regulation on the part of workers to be constantly available, visible and accountable to work institutions in a domestic setting.

Because an organisation is comprised of co-workers as well as employers, the potential intrusion of mobile communications in both work and domestic activities can be carried out not only by employers, but also by co-workers. Laurier (1999) describes a situation in which co-workers ask and respond to questions about where they are to achieve the practical organisation of their work routine, and also to establish common contexts for communication. I'd like to suggest that by communicating where they are and what they are doing, individuals can simultaneously monitor their own and each other's work practices. As Laurier notes in Chapter 4, the strategic management of communications – switching the phone off, switching to voice mail or using call screening – are technologically mediated and practical strategies by which individuals can manage time and space, and create connections amongst the dispersed individuals that comprise a mobile organisation. They are also strategies that can be employed to resist surveillance of work practices by others. On the one hand, according to de Gournay (1999):

> [p]reviously legitimate rules guiding individual conduct in interaction with others, such as the minimal obligation to be accessible to the public when one is engaged in contractual relationships (at work, in any professional service) have become obsolete with the use of mobile phones.

On the other hand, the extent of obsolescence of mutual surveillance and account-ability seems questionable. One means by which to think about whether this mutual surveillance has become normalised or not, however, is to imagine what might happen if people routinely *resisted* questions about where they were or what they were doing in mobile communications in everyday life. In a work communi-cation, the moral imperative to account for one's present location and activities might not be very strong, but is nevertheless implicated in a web of micro-level power relations. In response to the question "where are you?", one could probably get away with "it doesn't matter". Would the moral imperative to remain account-able, however, become stronger if one refused to account for why one's phone was switched off for long periods of time during the working day?

Similarly, this mutual monitoring and accountability could be examined in the context of interpersonal and intimate relations. Consider the social consequences of refusing to account for your whereabouts or activities to a friend or lover (or a parent in the case of teenagers) while in mobile communication. "It doesn't matter" might be socially acceptable with some further negotiation, but imagine the social consequences of "oh, I'm just around", or "it's not relevant", or, in the extreme, "I'm not going to tell you". Such possible resistances, and why one might or might not use them in the context of the relationship, map the boundaries of mutual monitoring and accountability in emerging, mobile relations. The extent and implications of these questions remain, as yet, unclear however. These propo-sitions certainly require further and more specific research.

3.4 Surveillance and Accountability

These different forms of social interaction via mobile technologies certainly suggest that the practices of mutual monitoring via information and communica-tion technologies are shifting in ways unaccounted for in many contemporary the-ories of "surveillance" and self-regulation via identity and community. This is the case when the "surveillance" capacities of mobile networks potentially allow wider, distributed, more intense and more comprehensive forms of institutional surveil-lance in both production and consumption.

The limits of the concept of surveillance to account for more widespread mutual monitoring amongst populations become clear when we consider increasing mutual monitoring of whereabouts and behaviour via mobile technologies amongst various groups of people.

Our research indicates that everyday understandings of what "surveillance" is and what it might mean in terms of everyday practice and experience are highly contextual. What parents define as concern for safety, for example, may be experi-enced by teenagers as "surveillance" of their activities, and teenagers often talk about the strategies they employ to resist that, sometimes using the devices them-selves as an excuse to resist parental monitoring ("the battery ran out").

Questions of "surveillance" and mutual monitoring, visibility and accountability raise issues of privacy and trust amongst individuals, organisations and institutions. It should be noted that the notion of privacy is both contextually and culturally/legally specific, and not always extensive enough in its implications to understand the quality or extent of potential change in emerging mobile relationships. While individuals have no choice as to the location information they provide simply by their use of devices, many individuals may well be happy to give up their abstract "right to privacy" in specific circumstances in order to use global positioning to find a cab on a dark, sleeting night in the middle of London. Similarly, the legal and cultural notion of "privacy" is certainly not the same in the UK, as it is in the USA, for example. Moreover, there is some suggestion that different social groups will understand notions of privacy and technologies of surveillance in very different ways (McGrail, 1999). All these factors influence social expectations of the boundaries between legitimate and illegitimate surveillance, acceptable or unacceptable mutual monitoring and accountability, and the characteristics that define privacy and mutual visibility and availability. Surveillance, regulation and mutual accountability are therefore sites of struggle and negotiation for identity, activity and control of them in everyday life.

This negotiation raises several issues in the design and organisation of information and communication technologies. Information and communication convergence, and the seamless connection of voice, text and multimedia might be questioned by those who want to use their phone as a phone, without the potential "surveillance capacities" of global positioning, or the institutional intrusiveness of unsolicited image enhanced advertising. The "killer apps" may well be those which are the most simple, and the least intrusive. Certainly, user control over the information they provide about themselves, whether technically or contractually maintained, will be an important issue in struggles of data privacy.

The degree to which users will be able to control the information (such as location) they provide (simply by their use) to organisations, and the information they thereby receive from them, has implications for the organisational and business models that are evolving around wireless applications. Who owns and controls information about individuals, and how that information is used in the context of organisational contractual arrangements and strategic alliances, will become more important as public debate about surveillance continues and the surveillance capabilities of mobile devices are addressed at a technical level. For these reasons alone, this issue is deserving of further research.

What is just as sociologically interesting, however, is the social normalisation of monitoring practices at the micro-level of everyday life in both public and domestic settings. Routine surveillance by institutions and organisations normalises relations of watching and the self-regulation that results. Current research suggests that individuals can use their mobile devices to assist in their own surveillance by institutions, as well as resist it. At the same time, they also engage in routine monitoring of themselves and each other through those same technologies, and assume that others are self-regulating and accountable for their use of devices in both co-present and tele-present contexts. While these mobile relations retain elements of both "surveillance" and resistance to it, more research is needed as to how the boundaries of legitimate and illegitimate monitoring and accountability come to be defined in everyday life, amongst whom, and how notions of "privacy", the public and the private, the individual/personal and the collective, might come to be redefined in the process.

Notes

1. The research on which this chapter is based is entitled 'The Socio-technical Shaping of Mobile Multimedia Personal Communications' (STEMPEC). This is a three year project funded through the DTI Foresight Link scheme, and is sponsored by the ESRC, four network operators (BT Cellnet, One2One, Orange and Vodafone), and a digital content provider, Granada Multimedia. The project investigates the production, regulation and consumption of mobile technologies through qualitative and ethnographic methods. The data discussed in this chapter has been drawn from detailed observational work and qualitative interviewing.
2. The practice of social science research in this process is at issue here - when 'we' document the lives of those living in contemporary societies we are also potentially (some might say necessarily) engaged in a framework that normalises information gathering that institutionalises the process. Our research is theoretically and politically implicated in these dynamics of surveillance.
3. There is considerable debate into the effectiveness of CCTV in this regard, and a number of studies have indicated that cctv does not, as it's proponents claim, reduce crime rates. instead, they argue, it displaces crime into areas not surveilled by CCTV (see McGrail in McGrail *et al.* for a discussion of the social, economic and political implications of CCTV.
4. The STEMPEC research this paper refers to was conducted in one South-Central London High School, and Two 'Green Belt' Sixth Form Colleges. The high school in South London has around 1700 students, with an age range from 11 to 18/19. The school serves several London boroughs with a mixed socio-economic and ethnic profile. According to a deputy-head, the gender ratio is around 45% girls and 55% boys. The ethnic composition of the students is around 40% anglo, 30% african or afro-caribbean, and 30% asian. The two Sixth Form Colleges in the 'Green Belt' cater to an age range from 16 to 19. In one of the colleges staff provided estimates of approximately 70% phone ownership amongst the students, a percentage they suggested was pretty much evenly balanced between males and females. The London school had banned a number of different mobile devices (amongst other goods) at school for all but the Sixth Form. The Sixth Form Colleges had banned mobile phones during lessons.
5. Our research with teens and mobile devices has indicated that it is somewhat problematic to use the very term "teenager" to describe a group of people with common behaviours, interests or understandings. The identities of those who are 13 are very different from those who are 19, as are their behaviours and interests. Furthermore, others may categorise them in different ways: it is just as common now to refer to older youth as "young adults", for example.

References

Bogard W (1996) *The Simulation of Surveillance: hypercontrol in telematic societies.* Cambridge: Cambridge University Press.

Bordo S (1989) The body and the reproduction of femininity: a feminist appropriation of Foucault. In Jaggar A, Bordo B (eds) *Gender/body/knowledge: feminist reconstructions of being and knowing.* New Brunswick, NJ: Rutgers University Press.

Bordo S (1993) *Unbearable weight: feminism western culture and the body.* Berkeley, CA: University of California Press.

De Gournay C (1999) *The mobile phone: pretence of intimacy perpetual contact workshop.* Rutgers University, December.

Dichter M and Burkhardt M (1996) Electronic interaction in the workplace: monitoring, retrieving and storing employee communications in the internet age. *The American Employment Law Council Fourth Annual Conference*, Asheville, North Carolina, October.

Foucault M (1973) *The birth of the clinic: an archeology of medical perception* (trans. Sheridan AM). London: Tavistock.

Foucault M (1977) *Power/knowledge: selected interviews and other writings 1972–1977* (Gordon C trans. ed.). New York: Pantheon.

Foucault M (1979) *Discipline and punish: the birth of the prison* (trans. Sheridan A). New York: Vintage Books.

Green S and Harvey P (1999) *Scaling Place and Networks: an ethnography of ICT 'innovation'*, in *Manchester*, Internet and Ethnography Conference, Hull, December.

Hill EJ, Hawkins AJ and Miller BC (1996) Work and family in the virtual office: perceived influences of mobile telework. *Family Relations* Vol. 45, pp. 293–301.

Laurier E (1999) Why people say where they are during mobile phone calls. http://jimmy.qmced.ac.uk/usr/cilaur/dynamic/S%26Swhere2.html

Ling R (1999) Restaurants, mobile phones and bad manners: new technology and the shifting of social boundaries. In Elstrom L (ed.) *Human factors in telecommunication*, 17th International Symposium, Copenhagen, Denmark, May, pp. 209–221.

Loader B (1997) *The governance of cyberspace*. London: Routledge.

Lyon D (1994) *The Electronic Eye: The Rise of Surveillance Society*, Cambridge: Polity.

Mason D, Button G, Lankshear G and Coates S (1997) Technology work and Surveillance: organisational goals, privacy and resistance. *Virtual Society Programme*. http://www.brunel.ac.uk/research/virtsoc/projects/ mason.htm

McCahill M (1996) Beyond Foucault: towards a contemporary theory of surveillance. *Surveillance, Closed Circuit Television and Social Control*, C. Norris, J. Moran and G. Armstrong (eds.) Brookfield: Ashgate Publishing.

McGrail B (1997) The virtual remake of high rise housing: electronic technology and social space. *Virtual Society Programme*. http://www.brunel.ac.uk/research/virtsoc/projects/mason.htm

Rule J (1973) *Private lives and public surveillance*. London: Allen Lane.

Sawicki J (1991) *Disciplining Foucault: feminism, power and the body*. New York: Routledge.

Spender D (1995) *Nattering on the net: women, power and cyberspace*. Melbourne: Spinifex.

Chapter 4
The Region as a Socio-technical Accomplishment of Mobile Workers

Eric Laurier, Department of Geography, University of Glasgow, Scotland

4.1 Introduction

> Predictability is an astonishing collective achievement. (Manning 1992: 4)

If you work in an office. If you go to the same place every day to work. If you do the same hours, Monday to Friday every week. Maybe you see the same faces, pick up your lunch from the same sandwich maker, drive the same route or take the same train to get there. When you're at work, you have the same conversations about what you did at the weekend, what you read in the paper that morning or what you saw on television last night. You fill in the same forms. You answer the phone with the same name. It's stable, yes. Predictable, yes. Inevitable, no.

There are a lot of "if's" about this "you" and they may not hold true for the reader-you. Whichever you you are, it's a picture of office life that does not seem that odd. Yet such routines, such orderly interactions really are remarkable. They become all the more remarkable when they are witnessed ordering more unusual places. Predictability in these unusual places is seldom expressed as a complaint and it is much more likely to be understood as an achievement. In sympathy with the general approach of actor-network theory (Law and Hassard, 1999) though more particularly with ethnomethodological studies of work I take predictability to arise not from "big" casual factors that precede (and might predict) situations of work but to be an everyday, mundane (and yet remarkable) accomplishment of those people and things involved (Garfinkel, 1967). These normally unnoticed features are rendered not just visible but bothersome when some of the "bits and pieces" so long associated with "the office", or rather that are the very stuff of associating the office, are used to assemble the office elsewhere. In this case I would like to describe some of the socio-material practices that allow offices to be achieved on the road, that un-fix or translate an office to make it mobile, diversely connectible and extensible. Offices, like laboratories, factories or law courts, seem to be "obligatory points of passage" in the circulations of people and things throughout modernity (Latour, 1987; Hetherington, 1997). They are at the very centre of human affairs by way of building long chains of translations that lead to them and away from them. Indeed, offices, are also "centres of calculation", to use another Latourian term. As Boden and Molotch (1994) put it, they are places that demonstrate the "compulsion of proximity", the need for organisations to have rich, unpredictable and intense encounterings to manufacture new orderings. Whether they are made of corridors and small rooms or open plan, hot-desked or romping places, their spaces are enacted through talking, walking, writing and a variety of other practices.

When we think about all these versions of "the office" we might imagine a big immobile building surrounded by other big buildings in a big city, like London, New York or Tokyo. Inside these offices we can observe "work in action" and begin to fill in the "missing what" that lies between the codings and summations of interviews and questionnaires administered to office workers and how such workers actually go about "doing" their tasks (Lynch, 1993). We can do ethnographies, blending the practices of the social sciences with those of experts in other fields to produce "wild" geographies and "wild" sociologies (Lynch, 1993). All too often socio-spatial investigations that claim to be "ethnographic" precede through an over-formalised, scientised process of interviewing, transcribing and coding while setting aside the "under-built" method of participant-observation (or what ethnomethodologists also call "hanging around with . . ." and office workers call "shadowing . . ."). To clarify this point what do we learn of office work by asking a manager what her job involves? In comparison what do we learn about office work by accompanying a manager as he does his job, even just for a few days or more? One "missing what" may be a corridor walk in its situated specifities as a manager produces her "overview" of the "state of play" of her employees through short sequenced chats, queries, updates, overhearings and overlookings (as observed by Mintzberg, 1973, Boden, 1994 and Hinchliffe, 2000, on "overlooking"). This is perhaps an over and oft-stated reminder to social researchers that ethnography's credo is to ask people what, why and how they do what they do *and* to compare that to what they actually do.

In this chapter I want to describe other offices cobbled together between their big brothers and sisters, decentred offices, displaced offices which have no corridors for managers to do their managing in (Boden, 1994) because these "lite" offices are travelling along motorway corridors. Via a somewhat strange rhetorical move in the social sciences I would like to deny that "the mobile office" is part of a *theory* or *explanation* of workplace geographies or that it is an advancement of Theory with a capital "T". Instead I will be offering a *description* of how one particular mobile office is assembled and organised; not as a case study for Theory to subsume by reductive analysis to yet another demonstration of how good theory is at explaining social and spatial order. Thus I will not be "stepping back" from the workplace I am interested in; in other words I not taking the "god's eye view". I want to show how mobility is organised "on the ground" by nomadic workers by bringing us, if not quite to the view from the driver's seat, then at least that of the passenger's.

Following Harvey Sack's gloss (see Garfinkel and Wieder, 1992: 186–187) I think I have found "a perspicuous setting" which provides us through "the haecceities of some local gang's work affairs, the organisational *thing* that [we] are up against" which is teaching us thereby what we need to know and learn about "work, mobility, technologies and regions". So to answer some long-standing geographical questions I joined a work group that faced these "organisational things" as ordinary problems which can be worked through with and in practical observable describable actions. "Mobile office" (like laboratory or factory) is a fruitful confusion of topic and resource and a shorthand for a local socio-material ordering of a situated practice which is constituted by highly mobile managers and related service sector workers who do a large part of their jobs in their cars, using mobile phones, Post-it Notes and roadmaps and other paraphernalia, whose "clients" are spatially dispersed. I think you might have a sense then of the "place" I am going to describe in what follows, though once again I have presented it to you in a kind of shorthand. "So let's get down to business."

4.2 Pre-assembled Parts

Although much of what follows will be concerned with the *assembling work* done by car-based mobile workers to keep their world relatively predictable, which is also an often unrecognised part of their occupations, many of their resources come *pre-assembled*, and their *stability* is not something to which the workers have to pay much concern, except perhaps as a constraint. This is to say that before any of them wake each morning, the highway system lies waiting, as does the associated signage, highway codes, service stations and so on. Outside their homes, their cars sit parked and ready to start – although this is not always the case since sometimes their cars are broken into, their wheels are stolen or the car engine fails to ignite. The extensive digital and analogue mobile phone networks await their connection – although again this is not always the case since sometimes the handset battery goes dead or the signal strength is too weak for connection.

Equally their business organisations remain relatively stable alloys of people and things, with their terms of employment, their multiplicity of contracts, their knowledge and their topologies (Grint, 1991; Thrift, 1994; Law, 1999). We might think of these pre-assembled parts as actor-networks: foldings of what is precarious and stable, far away and close, human and non-human, controllable and treacherous. They are effects at the more durable, predictable end of a socio-material ordering (Law, 1994: 139), which usually have limited flexibility, and as such, although we might be able to use a car as an office, we could not suddenly decide to use a petrol station as a fish farm. Flexing the pre-assembled parts into different and unanticipated uses requires work and resources in proportion to their orientation towards, and the fixity of other arrangements.

4.3 Assembling a Mobile Office

Sylvia[1] sometimes has as many as three alarm clocks set to make sure she wakes in time to beat the traffic on the highway into the city. So at 5.30am she quietly makes her way down the stairs, nips in and out of the bathroom, picks up the briefcase that she packed the night before and within 10 minutes she is in her car. Every morning when she uses the car for work (she tries to take the train when she can) she goes through the same four actions: slotting the phone into its hands-free holster; clipping the facia back on to the car radio and switching it on; putting her briefcase on to the back seat; and finally taking her shoes off (she likes to drive barefoot).

These may seem like quite banal ordinary activities, yet these are the *practices* that initiate the day's tasks, and these are the actions that re-assemble her mobile office. They have become routinised and efficient in the hands of Sylvia and the rest of the car-based workforce. They hardly need to be thought about, and yet if they were forgotten Sylvia could arrive at work without her briefcase or with cramps in her feet, or she could head off into the mêlée of inner-city traffic without her radio and realise that she had no current information on where the worst jams were and which of the roads in and around the motorway corridor were closed for repairs. Her world would become a little less stable and a little more unpredictable.

Having noted earlier that one of the pre-assembled parts that Sylvia could rely on was her "business organisation", I think it is worth re-examining such an entity here. A question we might ask is just how we might encounter such a "big" company. Her parent company is listed in the top ten transnational companies in the business directory that I consulted before doing this fieldwork with her. Yet take away the moment of arrival where Sylvia disappears into the lobby of a large building in the centre of London and what have we left? A question about what more specific attachments she has to the transnational company?

Classic workplace geographies of the kind done on assembly lines or in big offices produced a certain picture of what "big" and "powerful" meant and how such organisations succeeded or failed. Where work was, who the workers were and how surpluses were being extracted were relatively clear. Divergent accounts of work from feminisms, ethnomethodologies, cultural studies and more recently from actor-network theories have recast organisations less as entities (indeed their very claims to being any kind of causal agent are questioned) and more as processes of extension (Grint, 1991; Massey, 1994; Thrift, 1996). It may be worth flagging a concern which is not "ours" only as critical thinkers but is again an ordinary intelligible worry of Sylvia's: worrying over the *extension* and *intrusion* of her work into her "private life". Two of the ways this extension and intrusion was occurring was in the longer and later hours of her working day, and through her being on-call to her organisation and its clients via her mobile phone. This sense of a transnational company not as a big structure but as a loci of actions which are *organising specific connections* which can sometimes be extensions and sometimes be intrusions is all the clearer when someone's workplace is stretched across a region.

Except that it would be a mistake to grant "a region" some a priori spatial status; regions consist of concerted efforts, attempts, stretchings, associatings, descriptions and embodyings (Hinchliffe, 2000). They are not geographical realities waiting to be stretched across. Bringing about a successful region is after all one of the central tasks and accountable targets of travelling folk like sales reps, regional managers and area managers (as well as geographers). Such a (contingent) accomplishment is not necessarily attributable and is certainly not fruitfully understandable in terms of the stated intentions of each human region maker. In playing down *intentions* and *decisions* socio-material practices can be foregrounded as the activities that co-implicate a set of actors. Region-making happens in local associations of mundane activities, such as driving from client to client, typing in telephone numbers into carphones, etc., which are not "possessed by" nor usually even all that noticeable to those doing them, human or not; though one of the activities of making such practices orderly and distributable may be that they are able to be accounted for and recognised in some way and at some stage (Garfinkel, 1967).

So a story about "life on the road" that a nomadic office worker might tell would be about how they were "responsible" for the successes of their region because they spent a lot of time out of the office building bonds of loyalty with their clients, getting to know their "region" by ordinary practices that geographers, anthropologists and sociologists amongst others call "fieldwork". They can tell "war stories" (Orr, 1996) about their job and display the accumulation of local institutional knowledge gathered on the ground (even while apparently conversing about something else as we shall observe later in the chapter).

To return briefly to workplace geographies, if we accept that one of the modes of ordering which occurs in factory spaces (be they Fordist, Taylorist or Wedgwood) is the "assembly line" which brings about work through a particular channelling of materials and practices (the factory as a centre of flows) whose humans remain relatively immobile and parts are brought to them to be assembled, nomadic workers present us with quite a different spatial and socio-technical set of practices; they connect up their regions by constant mobility, assembly lines with moving humans as parts and relatively immobile materials (Law and Hetherington, 2000). A further differentiation of workplace geographies comes about through practices here and not necessarily because these workers are categorically defined as separate from factory floor workers.

If we regard the car as a mobile office, this helps us to visualise how what we think of as an office can be reconstructed elsewhere, but we do risk making too simple an equivalence between cars and mobile offices. My emphasis, to reiterate, is on "the assembling practices that associate diverse collections of people and things, and on the assembling of practices that these people and things then allow" (Hinchliffe, 2000).

Car-based workers flexibly flit from walking on the pavement to commuting on the train when the car becomes impractical or unnecessary. Sylvia takes the train to London sometimes three times a week and uses public transport. Office practices can migrate from the interior of a car to the interior of a train, to a city street, to a café. Indeed, it is a vital element of what they do that they connect up these diverse scenes. Sylvia has most of her promotional material mailed or emailed to her, and has "hot-desks" so that she can use (immobile) office buildings in Bristol, London and Birmingham. Also, and importantly, her own house serves as a repository for filing cabinets, colour desk-jet printers, pens, floppy disks, spare promotional materials, and so on, which means that she leaves a partially assembled office (and storage facility) for early morning, late evening or weekend, or even weekday periods working at home (yet she is *not* telecommuting).

In the early to mid-morning slot as mobile workers head toward their first meeting of the day, speeds are set by a mix of the driver's mode of comportment and the surrounding traffic flow. Sylvia has travelled the M4 motorway hundreds of times and during the early hours of the day travels it very fast. Accounting for her speeding, she suggests that the M4 is less heavily regulated by the traffic police because it is a "business corridor" and the need to keep it flowing smoothly and quickly outweighs the maintenance of the 70 mile an hour speed limit. Her previous diesel-driven car before the current racy Calibra wouldn't go much faster than 80. The week that I am with her in the car is in the autumn and as we watch the sunrise at about 7am she starts dealing with her voice mail. It is at this point that the car merges with the office more significantly. If we think back to a period only a decade or so ago, Sylvia could not have been working like this. She could have been working in other ways, like planning her day ahead in her head or reflecting back on her experiences of the job so far, but this morning she is already in touch with her clients and her team. There are messages from the night before, and messages from other people on her team who are also on the road and in the office at this early hour. In the morning in the car Sylvia is on-call, and so there are not only delayed messages but phone calls being received and phone calls being made. An important part of the work that goes on during this morning period is sorting out what needs to be responded to immediately from what can wait until later. There are socio-material mechanisms for this sorting: voice mail which, while allowing Sylvia's availability on the phone to be stretched out beyond her immediate ability to answer callers, also allows her to rank the messages she has been left in order of importance. Some may not need to be responded to by phone at all; they may be things she needs to know that day such as a change in company policy or a warning about some problems with their company sales figures. These little tips for the day frequently get noted down along with other messages on tiny yellow Post-it Notes that Sylvia sticks to the black airbag mound in the middle of her steering wheel.

The mobile phone itself gives off the appearance of a sleek self-contained portable object of the late twentieth century, and yet it is a relational entity par excellence constituted by its connection and disconnection to other people and things. It's a truism to say that a disconnected phone is a useless phone, though I would argue that "connecting" is not as on/off as opposing it to disconnection implies. For Sylvia in the above vignette is combining a form of storage known as voice mail, accessible through her mobile phone, which she is busy translating into return calls, brief notes on Post-it Notes, emotional clues for the day ahead and frequently "requests" to be noted and ignored. And where is her voice mail? Both distant and part of the local ordering in this and all cases and only distant if we take a naively Euclidean view of Sylvia's workplace (Law and Mol, 2000). In initiating the day's tasks Sylvia has delegated to her voice mail, a perhaps remarkably human action since it phones her up and initiates several conversational turns: firstly telling her that she has messages and then that she can receive them should she choose, and after she has listened to them offering her several other options such as immediate erasure, erasure after 24 hours or archiving. Voice mail allows for the possibility of responsiveness without being drawn into the particular "concertedness" of phone conversations. Dealing with voice mail is nevertheless something that is done in "real time"; it does not offer Sylvia or any other user time out of time. She can use it to "flick" through her messages. Doing this "flicking" still involves working her way through the turns that her voice mail provider has deemed are "normal workflow" for an answering service. Sometimes this means that in its step-by-step procedures it slows Sylvia down. For that reason she has deleted the archive option to spare her that additional question since she never bothers to archive her voice mail.

Voice mailing is connective since it allows her time-space availability to be extended, as well as dislocational since it orientates her toward distant and non-immediate requests and responses. So the lonely life on the road which used to typify the experience of "travelling salesmen" [*sic*.] can be partially refigured through the ability to interact with a team of co-workers without being corporeally co-present. It is worth noting that the various turn-storing devices, most significantly voice mail, are usually still directed towards "live" conversations rather than ends in themselves. Sylvia, in common with the other mobile office workers involved in the research, seldom left "completed" messages on voice mail. Her preference was to formulate a matter for a conversation to follow (often with a time she would be available "live"). A matter such as: "about the new sales promotion … can you call me back. I'll be in around 4.30". In this sense her voice mailings were used for greetings and a "reason for calling" which would help the person receiving the "mail" anticipate the "call" which was still thereby effectively waiting.

A constituent feature of work in and of a mobile office is this: with the need to *centralise* and to *intensify* relationships in the ordering that is a large business company, a need normally served by having a large building where teams can interact and create strong bonds of trust, collaborate on tasks and share out knowledge and skill, the activities of the mobile workers who create regional orderings for these companies are constantly *decentralising* and *dislocating* discourses and materials into the places of their clients. They are perhaps in an ethnomethodological sense remedying client's indexicals for their organisation. In Susan Leigh Star's (1992) terms, they are at work on the "extraordinarily complex and delicate task" of inhabiting "many different domains at once" (p. 52) creating bonds of trust, collaborating on tasks and doing knowledge work which runs along a border between (at least) two communities of practice (Wenger, 1998).

To maintain the organisational orderings, called "teamwork", at a regional scale requires the kind of *non*-face-to-face conversational skills that we have looked at above. It requires creating faces/identities for the team through the possibilities offered by telephonic interactions, *without* being able to gaze upon one another, a task that is made more difficult because of the usual dominance of visual cueing and in particular eye contacts. How does someone politely avoid "eye contact" over the phone and thereby emphasise their disattention, or how do they make direct extended eye contact and thereby emphasise something they are saying? (Such "moves" in talk are the staples of Harvey Sacks' (1992) investigations of a kind of local ordering that is telephone calling.) And just as actual face-to-face interaction with clients and team members requires several performances of faciality – situational propriety, involvement, accessibility, civil inattention, embarrassment (Manning, 1992) – so too does the face work over the phone.

When dealing with other team members a great deal of mediated "face work" is required since it is the office that has been stretched across space to allow for the greater quantity of co-present face work with clients (Hughes *et al.*, 1999). The situated nature of work means that there are a particular set of geographies of team performance in every setting since, as in the case of Hochschild's (1983) renowned work with airline cabin staff, the clients come into the *team's place*. In their staff quarters and inside the plane, the team know their way around the stage and the props, they have time together to review their past performances, to work-up new ones, to monitor one another in action and to "closely" co-ordinate "what to do next?" (Garfinkel, 1967). Other service industry investigations have also tended towards team settings into which the clients enter and so the service team has the added advantage of managing the stage on which the interaction will be played out (Goffman, 1956; Burton, 1994; Crang, 1997; Whyte, 1948; Ritzer, 1996).[2]

It's still early in the morning when Sylvia nips down into a tight-cornered underground car park. It's near a greasy spoon café where they serve mugs of caffe latte out of a battered old Gaggia. With a mug of tea or coffee comes a round of hot buttered toast. This is the place where Sylvia's crowd from her company take breakfast when they can. True to form a colleague of Sylvia's called Jane comes into the café after we've been there for five minutes. They talk about the times they left that morning: Jane caught a 6am flight to make it down in time for a meeting at this end of the country. Normally at this time she would be out on the road somewhere in the north. Today she's flown down to discuss "road-shows" which are going to motivate the transnational company's clients. She and Sylvia talk about this and about their worries that the company is about to make them take on a second client group much to the displeasure of their first client group. Jane has been called to London to make sure she takes and makes the "right impression" about this change in the organisation.

The vignette above tells a short tale of how a global company is brought together in a local situation, and it is about remedying some of the ills from which the transnational company suffers. Company HQ is only a couple of blocks away from the café and it offers "hot desks" rather than hot-buttered toast. It seems obvious where it is easier to run through company business in an informal mode (Boden, 1994). Jane is one of the TNC's "distant" employees since the main office in the UK is in London and her region is Scotland, Northern England and Northern Ireland. During periods of reorganisation from the "centre" Jane is called down for training weekends and mid-week meetings a great deal more. Her airplane commuter runs to attend face-to-face meetings are an additional element alongside the weekly video conferences and the daily phone calls she makes when she is out on the road crossing her region.

4.4 Boundary Work

As we approached Sylvia's first appointment of the day, Sylvia explained about knowing what was the appropriate time of the day to try to make various kinds of phone contacts; for instance the message she had left in Jane's voice mail a couple of minutes earlier – she knew it would be voice mail because Jane was always busy at that time. Equally Jane knew to phone Sylvia back at a certain time when she would be on the road again – in other words in her "office" again and not in a meeting with clients. Other tricks were calling people who you knew you didn't want to speak to directly at all – so you picked the times that you would get their voice mail. Because Sylvia's phone displayed known callers' names and numbers while the phone rang, she would in certain cases not pick up at all and let them leave voice mail instead. There were ways apparently that the organisation could check times when voice mail was left; and for how long it was left, Sylvia added ominously. Also travelling with Sylvia I noticed how long she spent just "trying to get hold of someone on the phone". Connecting to her clients on the phone is not a simple operation; it's a long series of hailings that may eventually get a response. Everyone is busy call screening everyone else. If they don't, then Sylvia wonders what's wrong with their business!

Call screening is a finely crafted skill among mobile office workers. Constantly being on-call to anyone who has your number is one of the dilemmas posed by the mobile phone, and call screening in its various forms is a way of dealing with this dilemma. Being on-call has various repercussions in terms of both surveillance by the organisation and the possibility of being swamped by client calls (the intrusions I noted earlier). One of the characteristics of an immobile office is that its architecture can be used to firm-up multiple boundaries, and thereby assist in workers' time management/availability. Entering the door of the organisation's building means being at work and being witnessed being at work as both a possibility of surveillance and display (Crang, 1994). At a different level, being away from the desk at a meeting means that a task can be completed without interruption because it is happening in another space (which is a way of being off-call to other requests for attention). Even in immobile offices, this other space in which one should not be interrupted is often in the same room, with either the phone diverted and/or a "meeting in progress" sign on the door. Nevertheless, with the different architecture provided by the car as a mobile office, further forms of boundary and demeanour work are required (Tolmie *et al.*, 1998). In the case of Sylvia and many of her other colleagues, this was by setting times of day when they would be officially *in*, and this was generally the early morning and early evening. For the rest of the day their time was given over to doing face-to-face service work with their clients. One of Sylvia's colleagues held on to her carphone (it was cabled into the dashboard so that it could not be moved as a mobile phone out of interior of the car) long after it was superceded by car-kit clip-in, clip-out mobile phones to make materially manifest where the "front-line" was between her office and her being out of the office – on-call and not on-call.

These timetables were informal and by their informality bonded and bounded team members who would know about when someone would be "in" or "out". Clients (and superiors) were less likely to be aware of Sylvia's "availability" unless specifically told to call at a certain time and even then it was not brought to their attention that this was the time-slot they could rely on reaching Sylvia. The informality further served the purposes of filtering who was allowed to draw Sylvia into conversations, since, by using her display to warn her of people she did not want to be called by, she could let them leave voice mail.

Certain interactional turns are delegated to the mobile phone system in a slightly less flexible way than to a PA or personal receptionist. Thus, various greeting interchanges which allow caller and called to identify themselves are done by the phone, acting as an electronic receptionist which matches a phone number to a name and affords Sylvia a degree of discretion in her "being on call" since she has an indicator of who is calling before they can be sure that she is available to them. For outgoing calls this means that the numbers stored by Sylvia are "screened" on an LCD on her mobile as names that she can flick through by pressing one button and then selecting to call these "names" with another – at which she point she may well be call screened by another human or non-human receptionist.

It is worth noting that this "memory" function of the mobile phone system, allows it to absorb (simulate/mimic) the material distinction of the paper address book. For incoming calls the caller's name, if it corresponds with a memorised number, is displayed by the phone, which is how the initial caller identification turn ("Hi, it's Sylvia di Maggio") is delegated to the phone system. "Hi it's Sylvia di Maggio" is an opening sequence worthy of investigation in its own right since a simple "hi" or "hello" does different kinds of conversational ordering. We might reasonably guess from "Hi, it's Sylvia di Maggio" that the called is not on close terms with the caller since she identifies herself with both her first name and surname. Her surname is doing institutional setting for her. This is not something that is restricted to call screening in a car of course. It is, however, a way in which a region is netted together, since known callers are divided from unknown callers by this technique (Laurier, 1999). Knowing who is calling means that Sylvia or any one else called can shape a response conversationally and emotionally before they pick up. Sylvia was particularly expert at launching informal conversations, greeting known callers as "hon" (honey), asking about something personal and demonstrating that she remembers them as an individual and thereby framing what followed as a more intimate, affective and trusting encounter. These conversational gestures are used with clients to build a sense of the organisation's engagement and knowledge of them; and at the same time with unknown callers, it initiates the phone conversation on a more typical polite level of response (i.e. "Hello, Sylvia di Maggio speaking").

Phone calls were also used in other ways to distribute and demonstrate organisational and regional knowledge, for instance in trying to define the identity of particular clients, as in the conversation below between Sylvia and Paul. Here, even in this truncated version of their conversation, they devote a great deal of effort to working out who exactly is the new representative of their client at "the Richmond" department store. Paul had said that the representative's name was "Vanessa". A name was not enough, however, and Paul begins a long series of turns at producing a biography of the client to see if he and Sylvia are agreeing on who they are talking about. And the complexity of this activity is suggested by the overlaps and the stops and starts of their turns.[3]

S: Was she by any chance like an assistant store manager at the same time
   Rachel was?
P: *Yes!*
S: Then [I *do* know].
        +
P: [in Cardiff].
S: Then I do know. [A skinny]
            +
P: [Yeah] I thought you would, yeah.
S: A skinny red-hair?

P: Naah::, naah::. Sort of mousy.
S: Yeah:::.
P: Yeah.
S: I think I *do* know her actually. [Talks a good story].
                                              +
P: [I mean, I mean she's in Cardiff], I think she moved to
    Exeter and then she moved to Glasgow.
S: She talks a good story but I don't know what she does:::.
P: Yeah, don't know. [I don't know].
                       +
S: [Mmhm].
S: Ahem 'cause she was in Exeter when it changed. It had a massive fashion
    refit.
P: That's it. Yeah.

After this final "yeah", Paul shifts on to another topic. What is worth noting is that it remains unclear whether Sylvia does know the same new representative as Paul – she fails on matching hair colour, comes close on "talks a good story" and seems to have finally shown she knows her with "Exeter" and "a fashion refit". It may also be that they have simply exhausted their interest and also reached an ordinary constraint on how many turns in a conversation can be devoted to making sure of someone's identity. Paul's double "don't know" on the fourth from last line suggests that he is closing his attempt to describe Vanessa and may be quitting the attempt with an agreement even if their versions of Vanessa still diverge. What we do see in this fairly long attempt at identifying "Vanessa" is a kind of question and answer game in reverse, which both Paul and Sylvia show themselves as competent at doing, with knowing when to stop the process being part of their competence. Indeed, showing their competence at doing this may be as important as actually coming to any conclusion about whether they are talking about the same person, as does showing a geographical knowledge of their client (Cardiff, Exeter and Glasgow) and what happens in those different places – Exeter "changed" and "had a massive fashion refit". They are not pursuing necessarily pursuing the "real Vanessa" here, something that becomes apparent if we concentrate on language as action (unfolding, co-ordinating and open-ended) rather than reducing it to language as representation (refers to Vanessa).

4.5 Out of a Mobile Office

Sylvia gets out of her car and steps inside a department store. She leaves me the keys to the car. I get out my journal and write some notes about Sylvia getting out of her car and how there's only so much of her job I can be involved in without becoming a nuisance.

When mobile workers leave their cars they are not only walking into their clients' places to have face-to-face meetings, they are also often joining up with the *materials* of their company which are residing in those places already. For Sylvia this meant that she would be inspecting the operation of her company's system *in situ*, and helping to settle it in with their clients. The system was not autonomous, and, like a technical maintenance crew for a photocopier, Sylvia and her team were supporting its operation. They performed emotional labours with the staff at the stores, sharing their disappointments, placating their anger, making their jobs seem "fun", encouraging them to be interested in sales promotions: in particular the bonus schemes. Aside from these more intangible affective tasks, they would also be reinforcing procedures to do with setting up new accounts, going over step-by-step instructions and chasing up stores that had lax security arrangements. Their tasks were of course not self-defined, and were done on site because the presence of their system machines (application forms, sales forms and promotional material) was necessary. Observing it in context was necessary for them to offer their assistance.

4.6 Broken Parts

> *In the middle of the afternoon in the middle of the city Sylvia's car has a flat tyre, but she takes it all in her stride. As she explains, it's not like the time she had to drive to a meeting with her broken windshield pushed out and the rain in her face. Having only ever owned a private car, I sigh and begin thinking about the hassles and dirt of changing the wheel. Meanwhile Sylvia is on the phone summoning the car leasing company's roadside repair staff, after which she sits contemplating for a minute before noticing that across the road there is a hairdresser's. She nips over and has a facial while she waits for the repair to be done.*

Much is made in research on social interaction of the importance of "repair work" which is carried out when social interaction "goes wrong" (i.e. there are mis-hearings and misunderstandings, evidence of abnormal behaviour, etc.). Mobile phone calls frequently require repair work of a technical kind (Goffman, 1981) since they rely only on an audible line of connection between participants in an encounter, and that audible line is highly susceptible to interruption and interference. Returning to the "on-call" problem mentioned earlier, participants in the project all told of how they could exploit these breakdowns in the means of communication at a distance to escape being constantly on-call, or to close a difficult conversation by claiming that the line was "breaking up". Or, in the above vignette where Sylvia suffers a mobility breakdown in the form of a genuine flat tyre, she steals some time from the organisation to have a facial. Much of the skill in working on the road is in taking advantage of these spare moments on the wing (De Certeau, 1984); though it should not be forgotten that Sylvia's facial can also be taken to be part of the "behind the scenes" preparation of self-presentation she has to do in order to meet her clients face to face.

> *During my fieldwork with Sylvia her passenger-side door is broken – it doesn't close properly and produces a deafening low-pitched whistle when she's driving over about 50 mph. It makes me wonder about sick building syndrome when your building is your car, with its noise levels, air quality and long seating times. Also the battery in Sylvia's laptop is no longer any use, it charges for about a minute and then goes flat, so mostly she just leaves it at home. She has thought about trying to track down an adaptor for a cigarette lighter slot, except her model of car*

like so many contemporary vehicles no longer caters for smokers and so there's no power source. There might be an adaptor kit but Sylvia knows how long it took her team to get phone kits installed into their vehicles as standard and is unwilling to begin the struggle to have another bit of her office translated into her car. There are too many bits and pieces of people and things involved in adding just that "small" extra part to her assemblage. There is a long list of other things needing fixing first: her mobile phone displays an error message on its LCD when it is docked into its slot beside the gear stick. Sylvia is unable to work out what it means and it hasn't caused any apparent disruption to her use of the phone so far but she reckons she will need to attend to its warning sometime soon. There is a modem inside her laptop which might not be working because she can't log on to the company server but she doesn't "know enough about computers" to know whether it's the modem or the software. Most of her weekly figures arrive on paper or by phone anyway and she stills phones in her totals at the end of the week.

For all the repair work that is constantly going on to keep things moving along together, many of her mobile office's parts are no longer or never were secure bits and pieces. Sylvia boasted that she knew much more about the workings of her mobile phone than most of her colleagues, while at the same time laughing at her ignorance of where to even check the oil levels in her car engine (though she did know where to put in the water for the windscreen spray). Most of her computer software was a mystery she had no wish to learn about (though again she was very good with spreadsheet packages and PowerPoint). It does not seem to matter that so much of the "machinery" is broken or under-utilised (by the engineer's definition of it) or its mechanisms incapable of being formalised by its users. In the ongoing flow of conduct the work gets done because it never requires a "god's eye view" nor an engineer's particular competencies nor a geographer's definition of a region. It requires an "economical" movement through and of the workplace in all its heterogeneous and site-specific details. If it were so reliant on the reliability of *all* of its parts then this "actor-network" would never hold together. And if there is something "new" for actor-network theory to learn from the users of technology then it is perhaps in the many ways that the "finished" projects of engineers, designers and company CEOs become performed as another reality of that fila-mental, socio-technical assembly (Mol, 1999).

4.7 Summing up a Region

It is all too easy in doing research on work and mobility to find oneself missing just what the work *is*, since both the social sciences and "the management" want to trade in generalities and have the privilege of the *overview* (Law, 2000);[4] an overview which, as Hinchliffe (2000) points out, is reliant on *overlooking* the details, in the sense of both looking from above and, for various reasons, pretending not to notice some of the things one has actually seen. It is also dependent on mobile workers (and social scientists) summing up the regions via the use of, amongst other for-malisations, spreadsheets carrying rules for the reporting on and calculation of the "state of the business" on a day-by-day, week-by-week and month-by-month, client-by-client, product-by-product, etc. basis. Alongside the documentary work there are also the quick chats, longer meetings, video conferences and work residential weekends where each worker tells "war stories", explains successes and failures and is instructed on "what the company wants done next".

While the summary view produced by these ways of recording is certainly germane to getting the work of the social sciences *as a profession* done and also that of "the management", it must not be confused with the practical, situated understanding and arrangement necessary for getting one's office to travel, or with designing appropriate artefacts, policy, health care, legal support, etc. for mobile workers. Remarks about technologies such as cars, mobile phones and WAP somehow causing work to be faster, more mobile and more connected-up tend to misrepresent the technologies and their users, glossing over how their spatio-temporal arrangement in use is just as much about slowing down, holding things in place and disconnecting.

In the vignettes offered in this chapter I have attempted to tease out some of the situated techniques used by Sylvia to organise her mobile office each and every day. To remind the reader:

- Sylvia getting out of her bed, into her car and the things that are *in place already* and the things she has to *carry* with her and *replace*.
- Sylvia cruising on the motorway early in the morning *sorting* her voice mail, putting it in order of relevancies to her planned route, client importance and other background matters.
- *Meeting face-to-face and informally* with a member of her team at a greasy spoon café in the city.
- Later in the morning, Sylvia making mobile phone calls to her team, who, along with Sylvia, will be *call screening*, and her clients (who are likely to be available at that time) to respond, pursue, set dates and times, etc.
- Sylvia and a co-worker talking on the phone, *doing some identifying* whilst also building company knowledge (who is who, where is where, etc.)
- Sylvia leaving the car (her mobile office) to enter her client's place, travelling the company's network to witness its products being correctly, incorrectly or otherwise used in particular ways on site, and to push for their maximal use.
- An everyday bit of road trouble – a flat tyre – and Sylvia calling for assistance and rescuing the time by having a facial.
- The various parts of the mobile office system broken or unused as a *good enough* (and why more?) *economy of distributed knowledge.*

There was a great deal more to office workers' lives on the road than the already complicated, sequentially organised and artful activities of Sylvia that I have described in these vignettes. What the work is varies tremendously and remains to be discovered since the occupations of mobile workers are diverse (from nurses to undertakers to credit sales) as are the regions they are expected to travel around. Passenger seats, where I spent a large quantity of time, are reasonably good places to learn about how regions are made and remade and how abstracted and reductive problems of social science (like global versus local, time-space compression, the predictability of social life, the mapping and bounding of distinct regions) are either relevant, irrelevant or plain nonsense in the specific situations they might otherwise claim to explain.

Acknowledgements

Thanks to Chris Philo for his collaboration on the socio-spatial and technical research project on cars and mobile phones, of which this chapter is part of the analysis. Also Ged Murtagh, Venetia Evergeti and Barry Brown. The project was undertaken with funding from the ESRC (Grant No R000222071). The ideas and views expressed in the chapter are those of the author and are not necessarily those of the funding body. A greater than usual debt is owed to the co-participants in this research project, most notably "Sylvia", from whom I have learnt a great deal about mobility, early morning starts and hospitality.

Notes

1. "Sylvia di Maggio" is not the real name of the mobile worker involved in this research and names throughout have been changed for the purposes of confidentiality.
2. An exception to this is Murtagh's description (Chapter 6) of the interactional troubles in railway carriages, and their solution through the use of gaze.
3. Square brackets with a plus sign on the line between indicate overlapping talk.
4. Note "the management" are just as likely to find themselves on the road and dealing with its contingencies as "the workers", and this distinction is being used to highlight why the details of life on the road are ignored. Sylvia, for instance, views herself as part of the management sector of her TNC.

References

Bingham N (1996) Object-ions: from technological determinism towards geographies of relations. *Environment and Planning D: Society and Space* 14, pp. 635–657.
Boden D (1994) *The business of talk*. Cambridge: Polity Press.
Boden D and Molotch HL (1994) The compulsion of proximity. In Friedland R and Boden D (eds) *NowHere: space, time and modernity*. Berkeley, CA: University of California Press.
Burton D (1994) *Financial services and the consumer*. London: Routledge.
Crang P (1994) It's showtime: on the workplace geographies of display in a restaurant in South East England. *Environment and Planning D: Society and Space* 12, pp. 675–704.
Crang P (1997) Performing the tourist product. In Rojek C and Urry J, *Touring cultures: transformations of travel and theory*. London: Routledge.
De Certeau M (1984) *The practice of everyday life*. London: University of California Press.
Garfinkel H (1967) *Studies in ethnomethodology*. Los Angeles, CA: University of California Press.
Garfinkel H (1992) Two incommensurable, asymmetrically alternate technologies of social analysis. In Watson G and Seiller RM (eds), *Text in context: contributions to ethnomethodology*. London: Sage.
Goffman E (1956) *The presentation of self in everyday life*. Edinburgh: Edinburgh University Press
Goffman E (1981) *Forms of talk*. Oxford: Blackwell.
Grint K (1991) *The sociology of work: an introduction*. Oxford: Polity Press.
Hinchliffe S (1996) Technology, power and space – the means and ends of geographies of technology. *Environment and Planning D: Society and Space* 14, pp. 659–682.
Hinchliffe S (2000) Entanglements: geographies of domination/resistance. In Philo C, Routledge P, Sharpe

J and Paddison R (eds), *Power and resistance: geographies of entanglement*. London: Routledge.

Hochschild A (1983) *The managed heart: the commercialization of human feeling*. Berkeley, CA: University of California Press.

Hughes J, O'Brien J, Randall D, Rouncefield M and Tolmie P (1999) *Virtual organisations and the customer: how "virtual organisations" deal with "real" customers*. http://www.comp.lancs.ac.uk/sociology/VSOC/YorkPaper.html

Latour B (1988) *The Pasteurization of France*. Cambridge, MA: Harvard University Press.

Latour B (1992) Where are the missing masses? The sociology of a few mundance artefacts. In Bijker WL and Law J (eds), *Shaping technology/building society*. London: MIT Press.

Laurier E (1999) Geographies of talk: "max left a message". *Area*, 30, pp. 35–48.

Laurier E and Philo C (1999) X-morphising: review essay of Bruno Latour's "Aramis, or the love technology". *Environment and Planning A*, 31.

Law J (1994) *Organizing modernity*. Oxford: Blackwell.

Law J (1999) After ANT: complexity, naming and topology. In Law J and Hassard J (eds), *Actor network theory and after*. Oxford: Basil Blackwell.

Law J (2000) *Economics as interference*. On-line paper, Centre for Sciences Studies, Lancaster University. http://www.comp.lancaster.ac.uk/sociology/soc034jl.html

Law J and Hassard J (eds) (1999) *Actor network theory and after*. Oxford: Basil Blackwell.

Law J and Hetherington K (2000) Materialities, Spatialities, Globalities. On-line paper, Centre for Science Studies, Lancaster University. http://www.lancaster.ac.uk/sociology/soc029jl.html

Law J and Mol A (2000) Situating Technoscience: an Inquiry into Spatialities. On-line paper, Centre for Science Studies, Lancaster University. http://www.comp.lancaster.ac.uk/sociology/soc052jl.html

Lomax H and Casey N (1998) Recording social life: reflexivity and video methodology. *Sociological Research Online*, 3, U3–U32.

Lury C (1998) *Prosthetic culture: photography, memory, identity*. London: Routledge.

Lynch M (1993) *Scientific practice and ordinary action: ethnomethodology and social studies of science*. Cambridge, Cambridge University Press.

Lynch M and Bogen D (1996) *The spectacle of history: speech, text and memory at the Iran-contra hearings*. London: Duke University Press.

Manning P (1992) *Erving Goffman and modern sociology, key contemporary thinkers*. Cambridge: Polity.

Massey D (1994) *Space, place and gender*. Cambridge: Polity.

Mintzberg H (1973) *The nature of managerial work*. New York: Harper & Row.

Orr JE (1996) *Talking about machines: an ethnography of a modern job*. London: Cornell University Press.

Ritzer G (1996) *The McDonaldization of society*. Thousand Oaks, CA: Pine Forge.

Sacks H (1992) *Lectures on conversation*, Vol. 2. Oxford: Blackwell.

Star SL (1992) Power, Technologies and the Phenomenology of Conventions: on being allergic to onions. *A Sociology of Monsters*. J. Law. London, Routledge: 26–56.

Thrift N (1994) *Spatial formations*. London: Sage.

Thrift N (1996) New urban eras and old technological fears: reconfiguring the goodwill of electronic things. *Urban Studies* 33, pp. 1463–1493.

Tolmie P, Hughes J, Rouncefield M and Sharrock W (1998) *Managing relationships – where the "virtual" meets the "real"*. Paper presented to the European Association for Social Studies of Technology, Edinburgh.

Wenger E (1998) *Communities of practice: learning, meaning and identity*. Cambridge: Cambridge University Press.

Whyte WF (1948) *Human relations in the restaurant industry*. New York: McGraw-Hill.

Chapter **5**

Mobile Communications in the Twenty-first Century City

Anthony M Townsend, Taub Urban Research Center, New York University, USA

5.1 Introduction: The Arrival of Mass Mobile Communications in the City

The internet received widespread attention from social scientists in the years following the release of Mosaic in 1993 and subsequent growth of the world wide web. Yet, the technologies with which humans communicate changed significantly in many other ways during the 1990s. The mass diffusion of inexpensive mobile communications technologies avoided scholarly attention, perhaps because it seemed pedestrian compared to the fantastic, nebulous depths of cyberspace. Yet the mobile telephone represented merely the first wave of a torrent of personal technologies that were leading to fundamental transformations in individuals' perceptions of self and the world, and consequently the way they collectively constructed that world.

Mobile communications devices profoundly affected cities as they were woven into the daily routines of urban inhabitants. But beyond the general neglect of mobile communications in social science research, practising urban designers and architects had only addressed these new technologies on a cosmetic level, such as the design and placement of antenna towers. Even in Scandinavian countries (where these technologies have penetrated more deeply in social and business networks than anywhere else) the challenges to conventional notions of public and private space had yet to stimulate a major response from environmental designers. Most importantly, widespread and fundamental transformations in the very nature of mobility in cities for the growing masses of wirelessly-connected inhabitants are being overlooked.

This chapter explores the ways that cities and new mobile communications technologies were influencing each other's development at the turn of the twentieth century. By then, mobile telephone subscribers had rapidly outstripped the number of Internet users in the developed world. Observers estimated that in Finland, the number of mobile phones had surpassed the number of wristwatches. In the new megacities of the developing world, mobile phone networks were rapidly surpassing the reach of archaic, under-developed wired networks. While Internet use remained limited to the few and well-educated, entrepreneurs in squatter communities brought telecommunications to the illiterate poor through leased mobile phones.

The diffusion of the mobile phone was among the fastest of any technology in history, and they rapidly supplanted conventional wired phones. By 1999 there were nearly 500 million mobile telephones in use throughout the world, which accounted for over one-third of the world's entire installed base of telephones. Leading nations for mobile telephone adoption were mainly concentrated in Europe and Asia (Table 1).

The scale of this phenomenon clearly rivaled that of the Internet, which could only claim 410 million users during November 2000 (NUA, 2001). Furthermore, the failure of so many Internet-only retailing enterprises in 1999 and 2000 suggested a need to return to the spatial context of communications in social science research. In retrospect, many of the complex, time-consuming sites of the World Wide Web seemed to have less lasting significance than a wireless site that could accurately and reliably deliver directions to the nearest pizza shop.

Given these indications that mobile telecommunications represented a far greater phenomenon than the desktop internet, and one that potentially had far greater direct impacts on use and navigation of urban space, how could an urbanist begin to think about approaching the study of this new technology? What tools and techniques are available, and what intellectual frameworks are useful? This section looks at three sets of ideas that provide some insight into this large and complex question: the mobile telephone as a spatial technology, the shaping of mobile technology by urban form, and the psychology of objects.

Table 5.1. Leading nations in mobile telephone adoption, 1999

Nation	Cell phones per 100 inhabitants
Finland	65.1
Hong Kong	63.6
Norway	61.8
Iceland	61.2
Sweden	58.2
Italy	52.9
Taiwan	52.2
Austria	51.4
South Korea	50.0
Denmark	49.5
Luxembourg	48.7
Portugal	46.8
United Kingdom	46.3
Israel	45.9
Netherlands	43.5
Japan	45.0
Ireland	44.7
Singapore	41.9
Switzerland	41.0
France	36.4

Source: Compiled by author from OECD publications.

5.1.1 The Mobile Telephone as a Spatial Technology

This chapter is concerned with the long-term evolution of urban societies as they struggle, change and evolve through the introduction of new communications technologies. Popular theories of the social consequences of new technologies paint utopian visions with a broad and undiscriminating brush (Graham and Marvin, 1996). Unfortunately, it is this simple determinism which dominates the public discourse on technological innovation, undermining our ability to see the complex interweaving between society and technology.

Urbanism is essentially a style or framework for thinking about the interaction between people and the technology of the built environment. The field maintains a long tradition of thinking about the city in spatial terms – where people and things are located, and how this is created and consequently shapes activity patterns (for example Mumford, 1961 or Jacobs, 1961). These intellectual traditions of city planning permit the exploration of the social consequences of new communications technologies from a unique perspective. While in the past these techniques have been useful for understanding transportation and construction technologies – profound transformers of space – they can and have been turned towards information and communications technologies.

Consider the telephone, an inherently spatial technology – its sole function is to allow communication at a distance. Its impacts on the spatial structure of cities, however, were complex and often contradictory. While manufacturing activities worked better on a single floor and demanded large tracts of inexpensive land (away from city cores), administrative and decision-making functions of corporate headquarters benefited from access to specialised producer services like law, accounting and advertising by locating in central cities. The telephone supported both types of reorganisation. Factories could decentralise into the countryside since orders and reports could be delivered and co-ordinated by telephone. Yet at the same time, it is inconceivable to imagine a modern high-rise office building without the telephone – the elevators could not support the number of messages travelling by courier from floor to floor every second, every minute, and every hour of the workday (Pool, 1973; Gottman, 1973).

To this analytical problem, the mobile phone adds the complication of freeing this communications capability from a fixed location in urban space. Mobile phones add an element of uncertainty about physical location to urban interactions. In a typical mobile telephone call, the first activity to follow exchange of greetings is establishing context. Location information, delivered in response to the spoken query "Where are you?" was a nearly universal part of this process.[1] According to a survey by the mobile operator OmniPoint in the north-eastern United States, as many as one-fifth of cell phone users were lying about their location when talking on a mobile phone (Chihara, 2000).

5.1.2 Technology Shaped by Urbanism

While technology has powerful impacts, it is also shaped by society and history. The spatial structure of cities has long been a powerful force driving the development of mobile communications technologies, and the two remain tightly inter-twined. Unlike the sudden breakthroughs that led to the rise of the web and mass internet use, Cellular telephony is not so much a new technology as a new idea for organizing existing technology on a larger scale. In fact, vehicle-based mobile tele-phones have been available, if not affordable, in the USA since 1946 when AT&T deployed systems in 25 cities. However, the technology did not become practical until the 1980s when metropolitan areas were subdivided into "cells", so that scarce radio spectrum could be reused in non-adjacent sectors. The term "cellular" is derived from the geometric structure of the antenna grid that links mobile tele-phone handsets into terrestrial telephone systems. A city or metropolitan area is divided into a grid of hexagons, or "cells", at the centre of which are placed trans-ceiver antennas, or "base stations". As a subscriber moves from cell to cell, the network switches the call off to the base station in the next cell. Cellular technol-ogy lets radio channels be reused in non-adjacent cells, thus greatly increasing the number of subscribers a system can support. This key innovation was a great improvement over earlier urban radio-telephone systems, which served an entire metropolitan area with a single antenna, limiting total capacity in any city to less than 500 users. The cellular system was first proposed at Bell Labs in the late 1940s, but was considered unmarketable and was shelved until the 1970s (National Academy of Sciences, 1997).

Urban geography has continued to influence the evolution of mobile telecom-munications technology. In the late 1980s, the high density of subscribers in large metropolitan regions such as New York and Los Angeles overwhelmed the capac-ity of the analogue cellular system, spurring the development of digital technolo-gies that utilised scarce radio spectrum six to ten times more efficiently than analogue systems (SRI, 1998).

5.1.3 The Psychology of Objects

While the interaction between urban form and mobile communications technolo-gies is a useful starting point, relying solely upon traditional urban methodologies in this analysis is sorely limiting. For as useful as these techniques are, they are pri-marily concerned with physical scales that are orders of magnitude higher than the actual point of intervention of mobile communications technologies.

A more useful approach is that used by psychoanalysts such as Csikszentmihalyi and Rochberg-Halton (1981) to understand how people understand and develop relationships with objects. In this view, even everyday household objects act as a symbol or sign that often is objective and entices similar reactions from a broad variety of people. This chapter will consider how mobile communications tech-nologies both cause and permit the reshaping of individual behaviour and self-image. This will provide a stronger basis from which to speculate on how these changes will aggregate to cause wider transformations of neighbourhoods, cities and regions.

5.1.4 Complexity, Communications and the Changing Nature of Cities

The difficulty remains, however, in developing a model by which technologically-enabled changes in personal interaction at the micro-scale can produce larger-scale changes in the behaviour of social and economic systems and the way they manipulate the city's physical structure. Unfortunately, urban theory does not offer many coherent models for extrapolating individual behaviour to explain system dynamics.

New thinking about the ways in which complex physical, biological and social systems behave has developed the concept of emergent properties. Traditional urban analysis often assigns agency to a city as a unit (e.g. "City A is busy", "City B is unfriendly"), even though these characteristics do not arise as the result of any single decision. Rather these behaviours naturally emerge from the interaction of many individuals in extremely complex and path-dependent ways. As Resnick (1994) explains, decentralised systems such as an ant colony or a flock of birds exhibit complex behaviour even though they lack a leader. A flock of birds can thus be simulated by ordering each bird to simply follow the one in front of it, with random environmental influences causing disturbances that ripple through the flock and cause it to turn or speed up. Cities can be understood as an emergent property of human social interactions.

This chapter's central argument is that new mobile communications technologies are fundamentally rewriting the spatial and temporal constraints of all manner of human communications – whether for work, family or recreation and entertainment. Cheap, flexible, always-available communications let decision-making and management of everyday life be increasingly decentralised, though as much or more co-ordinated than before. As a result of this decentralisation, the complexity of these systems becomes greater and therefore less predictable. In parallel, this decentralisation creates new interactions and potential interactions between individuals that are dramatically speeding the metabolism of urban systems, increasing capacity and efficiency. The real-time city, in which system conditions can be monitored and reacted to instantaneously, has arrived.

The consequences of such a decentralisation of the co-ordination of urban activities present a twofold threat to city planning. First, the broad diffusion of advanced time- and space-management made possible by these technologies may not result in enormous physical upheavals such as those associated with the automobile. Mobile communications technologies can reinforce the competitive advantage of central city business districts by making them more flexible and efficient for face-to-face interaction. Yet, the same technology may also make megalopolitan automobile-based urban sprawl manageable and liveable by squeezing more capacity from highway systems. This fact dramatically complicates emerging internal conflicts within the city planning on issues such as new urbanism and urban sprawl by undermining the existing technological space-time regimes that have both driven the trends and framed debate. For example, it is commonly assumed that high levels of traffic congestion can help raise support for less wasteful development patterns. However, mobile communications reduce much of the "cost" of congestion, since time spent wasted in traffic can be utilised for communications.

Second, and far more importantly, massive decentralisation of control and co-ordination of urban activities threatens the very foundations of city planning – a profession based on the notion that technicians operating from a centralised agency can make the best decisions on resource allocation and management and act upon these decisions on a citywide basis. Most planning tools intervene at a much higher level – yet the dynamics of systems for simulating decentralised systems (like the turtle-worlds of StarLogo discussed later in this chapter) suggest that overall system behaviour is determined at a very low level by the specific nature of interactions between individuals.

5.2 Object and Self – Personal Relationships with Mobile Telephones

This section examines the powerful relationships that form between individuals and their mobile phones. Beginning with the first contact through advertising imagery, and progressing through novelty, assimilation and dependency, this section traces a process of adoption and adaptation to the constraints and capabilities of this new technology at the individual level.

5.2.1 Images of Mobile Phones in Advertising

Marketing and advertising images were powerful forces shaping how new technologies are ultimately used and perceived. These images played an important role not only in influencing, but also reflecting individual's relationships with new devices. The connotation that telecommunications provides an effective means to overcome spatial constraints on lifestyle was traditional motif in this industry. Thus, it was not surprising that advertising images of the mobile phone, whose unique defining quality is its detachment from visible tethers to the physical telephone network, are heavily slanted in this direction. Promotional images used by major wireless carriers are typical of the way the conquest of space through mobile telephones is romanticised. In one example, Bell Atlantic Mobile has used a photo of a woman literally standing on North America, mastering her domain through the use of a nearly unseen mobile phone (see also Chapter 11 for a discussion of the advertising of mobility).

Gender plays a prominent role in advertising strategies. For women, the desire to own a mobile phone is often portrayed as stemming from security concerns or the need to sustain roving ties to friends and family. For example, AT&T's promotions presented the mobile phone as a woman's best friend, helping her survive in a drab, grey and annoyingly persistent city.

Marketing directed at men, however portrays the mobile as a tool of fashion, power and virility (Katz, 1999). One of the most striking examples of this advertising motif appeared in the UK. To launch Virgin Mobile, Branson's brand new UK wireless power play, he appealed to Brits' basest instincts.

> Branson and a throng of nude models were seen cavorting around on the back of
> a flat-bed truck with the racy marketing slogan – "Pick me up, Turn me on, Use
> me to your heart's content" – emblazoned across the vehicle as it snaked around
> London's West End streets (Quigley, 1999).

Thus for women, the phone is sold as a security blanket for the uncertainty of
the city, while for men the seductiveness, anonymity and sexual variety of the city
are the primary lifestyle characteristics associated with the phone. Alcohol and
automobile commercials often use the same level of raw sexual imagery, yet there
is a personal quality about the mobile phone – a device that claims to eliminate the
fundamental human anxiety about loneliness – that makes these images particularly powerful. If men buy Branson's phone, will nude women be chasing them (or
calling them) day and night? While such speculation seems extreme, it is naïve to
assume any other intent on the part of the advertiser. Yet this association of technology and technological artefact with sexuality and virility offers the opportunity
to consider the next stage of man–machine symbiosis: the mobile phone as an
extension of the body.

5.2.2 Relationships with the Object: The Mobile Phone as an Extension of the Body

Individuals developed very intense emotional relationships with mobile telephones that are manifested and displayed in a variety of ways. For example, in the
late 1990s, mobile telephones replaced umbrellas as the most commonly left-
behind item on the trains of the London Underground. Ironically, however, there is
no mobile service in the Underground, and presumably people were holding their
phones for some other, less instrumental reason than placing a call.

This relationship between person and mobile phone is full of contradictions.
Personalisation of phones is an extremely popular activity: by entering commonly
called numbers, adding new songs to replace the standard ring alert, or buying
colourful clip-on faceplates to replace the standard black matte. Yet the subscriber
information module (SIM) which stores all of the data necessary to tell a phone
what its number is and who owns it, simultaneously makes the object of the mobile
phone replaceable in an instant. All personal information is contained within a
small smart card that can be dropped into any phone chosen on the basis of cost,
technical sophistication or cosmetic appearance. The phone is a commodity, yet
the information coordinates of the telephone network that it represents have a
powerful pull. These coordinates have themselves become important status-
bearing personal identifiers.[2]

Despite these conflicting trends in personalisation and depersonalisation of the
physical artefact, the mobile phone quickly became perceived by users as an extension of the body, though perhaps more in an emotional sense than a purely physical one. As *Wired* magazine reported in 1999:

> In the last couple of years, Finnish teenagers have quit referring to their mobile
> phones as *jupinalle* – "yuppie teddy bears" – started calling them *kännyka* or

kanny, a Nokia trademark that passed into generic parlance and means an extension of the hand. (Silberman, 1999)

From a very physical metaphor of something separate, cute and essentially useless – a "teddy bear" – the name evolved into a more abstract, loaded term: a metaphysical extension of the hand, a serious tool linked to the owner on the most basic level. And this deep cognitive link between the phone and the owner was a persistent theme. Katz reports a study in which:

> Relative to either home or work phones, [digital mobile phone service] was judged to be much like a friend; one user even said that the [mobile] handset began to feel like a "part of my anatomy". (Kiesler *et al.*, 1994 in Katz, 1999)

In Japan, one of the highest rates of mobile phone penetration in the world has led to the introduction of mobile-aware jewellery. An Osaka-based company developed an artificial fingernail featuring a tiny light-emitting diode which glowed red or blue when the user's mobile phone activated. The nail could be filed to any shape and it required no batteries, since it was powered by the mobile's wireless transmissions (Boyd, 2000). The fashion and apparel industries recognised this trend early on and have incorporated pockets to accommodate mobile phones in bags, purses, trousers, jackets and belts.

5.2.3 From "Cellular" to "Mobile"

The personalisation of mobile telephones was also evident in the evolution of their names in English-speaking countries. In the 1980s and early 1990s the commonly accepted name for these devices was the wireless or cellular telephone. Wireless differentiated these new devices from traditional wireline phones, while cellular was derived from the geometric structure of the antenna grid that linked these devices into terrestrial telephone systems.

The mass diffusion of mobile telephones in industrialised nations during the second half of the 1990s, coincided with a decisive shift away from the cellular designation towards the use of the term mobile telephone. This indicated a broad shift in cultural perceptions and marketing campaigns from a view where the technological innovation was seen to be in the supporting infrastructure (cellular) to one where the intelligence is embodied in the device itself (mobile). And unlike linking oneself in one's mind to some complex and constraining grid of antennas, the idea of augmenting oneself with a tiny, smart device was very appealing.

5.2.4 Cutting the Cord: The Telepathic Future

As the 1990s proceeded, popular culture in nations with high numbers of mobile phone users began accepting and reflecting this perception of the mobile phone as an extension of the body. In a *New York Times* article a graphic designer, who produced signs for restaurants to discourage the use of cell phones, pondered this link:

> A cell phone is "a pacifier for adults", said Maira Kalman, the president of M&Co,
> a Manhattan product and graphic design group. "It makes you feel connected, that
> you're not alone on this planet." (Louis, 1999)

From pacifier, to nipple, to digital umbilical cord, the mobile phone rapidly progressed to assume a vital place in the virtual biology of urban information societies of the late twentieth century. At the final extreme, the mobile phone's connectivity might be completely subsumed into the body, and all other forms of communication become redundant – email, web, phone calls, all can be delivered over the universal handheld. People became dependent upon the connectivity that the mobile telephone represented, restructured their lives and personal habits around the device, and found it impossible to go back. Accustomed to the flexibility of scheduling, the freedom from punctuality permitted by constant ability to update others, it was nearly inconceivable to live any other way.

Instead, many people opted to give up their conventional wired telephones rather than their mobile phones. In 2000, as many as 2% of American cell phone subscribers were estimated to have "cut the cord".

The consequences of large numbers of people cutting the cord are enormous. Telephone numbers no longer become associated with places, such as an office or home, but with individual people. Those who "cut the cord" forego the option of not answering the phone in many circumstances, as there are few excuses when it is public knowledge that they rely completely on the mobile phone for connectivity.

The most important change that occurs in observations of subjects who completely adapt to the new lifestyle opportunities of mobile phones, however, is that time becomes a commodity that is bought, sold and traded over the phone. The old schedule of minutes, hours, days and weeks becomes shattered into a constant stream of negotiations and rescheduling. One can be interrupted or interrupt friends and colleagues at any time. Individuals live in this phonespace – they can never let it go, because it is their primary link to the temporally, spatially fragmented network of friends and colleagues they have constructed for themselves. It has become their new umbilical cord, pulling the information society's digital infrastructure into their very bodies. In fact, as technological evangelists at Nokia pondered, mobile communications could eventually evolve into an activity almost indistinguishable from telepathy (Silberman, 1999).

5.3 City and System

As the preceding section of this chapter has shown, the mobile phone has tremendous influence over individual's self-perception. Yet, the goal of this chapter is to assess the way in which mobile communications devices, the mobile telephone in particular, influence the evolution and dynamics of urban physical, social and economic systems. This section presents another set of ideas that can be used to frame thinking about how cities of mobile communicators might exhibit new behaviour at the system-wide level.

The first useful concept is that of the real-time system. Real-time systems are defined by an ability to constantly monitor environmental conditions vital to the operation of the system, and alter inputs to the system to achieve a desired equilibrium state. For example, chemical plants employ real-time systems to vent by-products from the reactors and keep reactions supplied with the proper chemical ingredients. Autopilot systems in commercial airliners rely on real-time sensing of the aircraft's movement and control surfaces to steer. Put simply, real-time systems operate by using feedback from sensors in one part of the system to either induce or inhibit activity in another part of the system, thus pushing it towards an optimum stable state chosen by the designer.

Economic and social systems work in a similar way. Prices are a form of economic feedback that directs producers and consumers to adjust their behaviour based on the availability of resources. Urban economies and social networks behave in a similar fashion. However, because of traditional barriers to the transmission of information – distance, time and power structures – urban systems have never operated in a real-time fashion. Significant delays in the propagation of information still exist, resulting in inefficiencies and delays between cause and effect.

In the 1990s, however, as mobiles slowly diffused through the more dynamic segments of the business population and urban professionals, some individuals and sub-cultures strived to achieve a real-time lifestyle. Yet, as Silberman wrote in *Wired*, it was no longer just yuppies that were changing the way they roamed and explored the city. In Helsinki:

> Ubiquitous [mobile phones] ... have transformed the way young Finns roam the city. They're taken a feature introduced by Nokia in 1993 – Short Message Service (SMS), a form of email you can send from phone to phone – and turned it into their primary means of mobile communications. Like schools of fish, kids navigate on currents of whim – from the Modesty coffee bar to the Forum mall for a slice of pizza or a movie to a spontaneous gathering on a street corner, or to a party, where SMS messages dispatched on the phones summon other kids or send the group swimming somewhere else. (Silberman, 1999)

And as Kopomma explained, this new nomadism served a very clear purpose in the real-time city – social survival:

> Life in a modern urban city is a social struggle rather than a struggle for physical survival, yet in this role, postmodern humans are every bit as nomadic as their Paleolithic ancestors. (Kopomaa, 1999)

This new mobility was the fulfillment of latent demand. The technology for such flexibility of movement had existed for decades – the automobile in particular. However, it was the diffusion of mass mobile communications, and the ability to co-ordinate individual actions and movements in real-time, that made this new urban lifestyle sustainable. The mobile phone broke the flow of information away from the scheduling necessary to ensure co-ordination of journeys. Information could be updated in real-time, negating the need to plan anything. Accessibility became more important than mobility. This new reconfiguration of time and space permitted to networks of friends and acquaintances using mobile telephones, became most clearly visible in the lives of people whose work required them to be on the go and in touch with a wide range of friends and colleagues, such as taxi drivers and teenagers.

Thus, the mobile telephone facilitated a huge explosion in person-to-person messaging, both through voice calls and interim technologies like text-based Short Messaging Service (SMS). What can be expected when these transformations are occurring throughout every system operating in the city, not just one isolated system? While it was almost impossible to predict what types of new behaviours could emerge, it was nonetheless important for urban planners to begin thinking about how these technologies were reshaping our understanding of the city at a fundamental level.

The major impact of mobile messaging and voice communications stemmed from the rise of the real-time functioning of urban systems. For as urban social and economic systems approached real-time operation, the metabolism of these systems increased. Drivers had better information on traffic conditions, and could plan alternative routes. The net effect was a maximisation of capacity utilisation of the roadway system. Extra flexibility in scheduling led to less peak usage of infrastructure systems, and more even, efficient spread throughout the day. Flexibility also meant that firms and individuals could commit fewer resources to tasks, and have a better capability to redirect or redeploy them as conditions changed or more information became available.

Thus, the main effect of mobile communications was a fundamental change in the traditional interaction between communications, face-to-face interaction, and large-scale urban structure that have determined the shape of cities since the early twentieth century. As Deutch (1977) wrote:

> The facilities of the metropolis for transport and communication are the equivalent of the switchboard. The units of commitment are not necessarily telephone calls but more often face-to-face meetings and transactions. For any participant to enter into one transaction usually will exclude other transactions. Every transaction thus implies a commitment. The facilities available for making choices and commitments will then limit the useful size of a metropolis.

With the mobile phone, however, face-to-face meetings no longer implied a fixed commitment. The lack of spatial and temporal constraints on the transmission of information through the mobile communications system resulted in a constant reallocation of resources, people and commitments to their most productive uses. The general result was an intensification of the use of capital over time, or a dramatic increase in urban metabolism. As mobile devices became more geographically aware of user location through radio triangulation and Global Position System technology, the addition of location-sensitive information added a whole new level of improvement to this on-the-fly decision-making.

5.4 Conclusions: Rethinking the City

This investigation of the diffusion of mobile communications in cities of the late

1990s suggests that a major re-examination of the technologically constructed nature of space and time in everyday urban life is needed. The introduction of a device that provokes such strong longings and desires in people, diffused rapidly throughout the entire population, and so fundamentally strengthened co-ordinated, decentralised networks of individuals, and seriously compromised the ability of urban theory to provide a convincing explanation of how cities function and grow. Furthermore, the naïve, even disdainful attitudes of city planners towards emotionally charged technologies in the past (such as the automobile) continued to push the profession to the fringes of irrelevancy. It could not afford to make that mistake again with respect to the telecommunications revolution.

5.4.1 Decentralisation ...

Personal technologies like the mobile telephone have their main impact on the dynamics of interpersonal interactions at a micro-economic level. Sociologists, like Castells (1996) have argued that the end of the twentieth century was an era of dramatic transformation where new institutional and organisational structures were emerging as a result of major technological changes in the economics of interpersonal interactions. Social, economic and political change rises from these powerful forces, acting at ever-more decentralised levels. This is in contrast to past eras when major technological innovations impacted the relations between states, corporations and other large institutions.

In city planning, however, the individual is rarely the primary unit of analysis. While Mitchell (1999) notes that "electronically serviced space for information work does not have to be concentrated in large contiguous chunks, like the commercial and industrial zones of today's cities", planners resist such a decentralising philosophy and remain focused on large-scale physical transformation. As a result, the widespread bit-by-bit reconstruction of cities is going largely unnoticed by a profession trained to visualising cities through aerial photographs, rather than streetscapes. Yet while we no longer see massive changes in land use that accompanied earlier revolutions in production, such as the growth of cities at the onset of the Industrial Age, profound changes in the use of urban physical space are occurring as a result of the real-time co-ordination of social and economic networks permitted through mobile communications. In a sense, the metabolism of urban spaces is increasing, as highly effective distributed communications networks replace bureaucratic, centralised ones.

The fundamental obstacle to understanding this micro-scale process of urban transformation is what Resnick (1994) refers to as "centralised thinking". The leaderless nature of decentralised systems is non-intuitive, often leading to mistakes in identifying them and understanding their behaviour. People assume that a flock of birds or school of fish must necessarily have a leader, when the group's behaviour is actually determined by a far more complex evolutionary process largely determined by the many interactions of individual neighbours over time.

Experiments with StarLogo, a software package developed at MIT to make the simulation of decentralised systems more accessible, offer an interesting way to think about the significance of individual-level technological intervention (such as mobile phones) on larger-scale social systems such as cities. In the StarLogo environment, each individual is represented by a "turtle" – a simple computational object that can move about the screen, interact with its environment, and interact with other turtles. By programming each turtle with a few simple rules about how to behave, it is fairly simple to achieve systems that exhibit very sophisticated behaviour yet lack "leaders"- the flocking of birds is one example.

Applying this model to cities, one can envision a set of rules governing the typical urban dweller's transactional exchanges with other individuals, and in fact this is an approach often used in micro-scale urban modelling techniques such as cellular automata. Prior to the introduction of the mobile phone, one would check messages at certain intervals, and then act or react based upon the received information. Communications, particularly for the physically mobile members of society, came in discrete bursts, and had to be tightly scheduled and planned to ensure that both parties were at their terminals to complete the call. With the introduction of mobile phones, however, communications become fundamentally different – as in real-time systems there is a continuous monitoring of environmental variables and returning feedback to the system.

Extending the StarLogo analogy, it is as if mobile phones are enabling a massive experiment of reprogramming the basic rules of interaction for urban inhabitants. The introduction of powerful and fundamental different modes of communication at such a basic level will certainly produce major transformations in large-scale behaviour of the system. It is interesting that the Finnish phone company Nokia's technology evangelists see the realisation of "practical telepathy" one day as a result of microscopic mobile phones implantable under the skin. While this is certainly a stretch of the imagination, that it is even a conceivable metaphor (and a convincing one) underscores the degree of shift in the temporal dynamics of telecommunication.

5.4.2 ... and the Postmodern City

The decentralisation and fragmentation of social communications, however, is a process that began in the 1980s with telecommunications deregulation and exploded in the 1990s through the Internet. Urban theorists such as Dear (1996) argue that postmodern urbanism was particularly characterised by such fragmentation, as large group identities based on common heritage and nationality decomposed along almost tribal alignments. The mobile phone certainly reinforced these patterns – it substituted chaotic, ad hoc decentralised networks for centralised ones.

Yet the mobile phone did not just arrive by accident. One school of thought, the social construction of technology, suggests that societies largely choose which technologies to develop based on current values and goals. As Casson (1910) wrote:

> No invention has been more timely than the telephone. It arrived at the exact period when it was needed for the organisation of great cities and the unification of nations.

Similarly, in the 1990s it seems that the mobile telephone arrived at the just the time when it was needed to facilitate dramatic decentralisation of communications channels required by new social systems in the postmodern age. In fact, the mobile phone is so well-designed for this task that Kopomaa (1999) has even called it a "postmodern form of communication". Particularly since the idea of cellular telephony had been around, but undeveloped, since the 1940s at Bell Labs, it might be argued that postmodern society chose the mobile phone for development during the 1990s.

Harvey (1990) argues that the conquest of space through time (faster travel, cheaper telecommunications) is directly linked to the condition of postmodernity, yet space and minute spatial variations between places remain as important as ever. The use of mobile phones offers an ever-finer level of identifying and exploiting minute variations in conditions between locations. Users micro-manage space by micro-managing time by always remaining accessible. And this use of mobile phones, to support tribal, nomadic, fragmented groupings is driving the reality of the city away from the rational, modernist regime that drives most land use planning.

At its core, postmodernism simply implies the inability of modernism to sustain order and coherence in the face of the acceleration of time and the compression of space in the twentieth century. Yet, mobile phones greatly improve the individual's ability to manage these forces, even to manipulate them in the way that large institutions have for decades. The use of mobile phones by protesters during the riots surrounding the 1999 World Trade Organisation meeting in Seattle starkly illustrates the potential. By being able to shift resources to flashpoints on the city streets faster than the local police, who relied on centralised systems for communications and decision-making, the opposition was able to gain a decisive advantage.

For urban planning, the mass diffusion of mobile communications may mean that the city is now changing far faster than the ability to understand it from a centralised perspective, let alone formulate plans and policies that will have the desired outcomes. Despite Batty's (1995) assertion that planners now have the technological capability to gather real-time data about the city – "the computable city" as he calls it – will the centralised view of cities that dominates planning impair understanding of this data? As Resnick (1994) notes, even trained observers make frequent mistakes in attributing intelligence when analysing complex decentralised systems. Without good models for understanding urban systems, no amount of data will produce useful analysis for guiding real-world decisions. Without a concerted effort to develop new knowledge and tools for understanding the implications of these new technologies, city planners run the risk of losing touch with the reality of city streets.

Acknowledgements

This chapter is a revised version of an article that first appeared as "Life in the Real-Time City: Mobile Telephones and Urban Metabolism" in the *Journal of Urban Technology*, 7 (2) 85-104. 2000. It is based on work supported by the National

Science Foundation under Grant No. SBR-9817778, "Information Technology and the Future of Urban Environments" (http://www.informationcity.org).

Notes

1. Verbal exchange is often a highly inefficient means of establishing context, and offers no assistance to those who may be near to the person speaking on the mobile telephone. Witzgall and Kaye (2001) suggest that mobile communications devices incorporate features that can deliver context information to by standers. For example, displays may indicate the likelihood that a speaker will tolerate interruption. This might be high for a causal business call, very low for a personal emergency call.
2. In New York City, following the telephone city code proliferation of the 1990s, a designer perfume "212" bore the name of Manhattan's original city code.

References

Adams R and Snaghera S (1999) Londoners losing track of phones. *Financial Times*, September 27, 1999.

Batty M (1995) The computable city. Keynote address – *Fourth International Conference on Computers in Urban Planning and Urban Management*, Melbourne, Australia, http://www.geog.buffalo.edu/Geo666/batty/melbourne.html

Boyd J (2000) Flashing the finger in Japan. *The Industry Standard*, April 3, 2000, p. 45.

Casson HN (1910) *The history of the telephone*. Project Gutenberg e-text, ftp://ibiblio.org/pub/docs/books/gutenberg/etext97/thott10.txt.

Castells M (1996) *The rise of the network society*, Vol. I. Cambridge, MA: Blackwell.

Chihara M (2000) Lying on the go. *Boston Phoenix*, March 18.

Csikszentmihalyi M and Rochberg-Halton E (1981) *The meaning of things: domestic symbols and the self*. New York: Cambridge University Press.

Dear M (1996) Intentionality and urbanism in Los Angeles, 1971–1991. In Scott A and Soja E (eds), *The city: Los Angeles and urban theory at the end of the twentieth century*. Los Angeles, CA: University of California Press.

Deutch KW (1977) On social communication and the metropolis. In Arnold B (ed.), *Urban communication: survival in the city*. Cambridge, MA: Winthrop.

Gottman J (1973) Megalopolis and antipolis: the telephone and the structure of the city. In I Pool (ed.) *The Social impact of the telephone*. Cambridge, MA: MIT Press.

Graham S and Marvin S (1996) *Telecommunications and the city: electronic spaces, urban places*. New York: Routledge.

Hafner K (2000) Hi Mom, Hi Dad. At the beep, leave a message. *New York Times* March 15, 2000. http://www.nytimes.com/library/tech/00/03/circuits/articles/16teth.html

Harvey D (1990) *The condition of postmodernity*. Cambridge, MA: Blackwell.

International Data Corporation (2000) *Wireless access to the Internet, 1999: Everybody's Doin' It*. IDC report unnumbered.

Jacobs J (1961) *The death and life of great American cities*. New York: Random House.

Katz JE (1999) *Connections: social and cultural studies of the telephone in American life*. New Brunswick, NJ: Transaction.

Kopomaa T (1999) Speaking mobile: intensified everyday life, condensed city – observations on the meaning and public use of mobile phones in Helsinki. Paper presented at *Cities in the Global Information Society Conference*, Newcastle upon Tyne, UK.

Louis E (1999) If the phone had a cord, you could strangle the user. *New York Times*, September 30.

Lynch K (1960) *The image of the city*. Cambridge, MA: MIT Press.

Mitchell WJ (1999) *E-topia: urban life, Jim, but not as we know it*. Cambridge, MA: MIT Press.

Moss ML and Townsend AM (1999) How telecommunications systems are transforming urban spaces.

In JO Wheeler and Y Aoyama (eds), *Fractured geographies: cities in the telecommunications age*. New York: Routledge.

Mumford L (1961) *The city in history: its origins, its transformations, and its prospects*. New York: Harcourt.

NUA (2001) "How many online?" http://www.nua.ie/surveys/how_many_online/n_america.html

Pool I (ed.) (1973) *The social impact of the telephone*. Cambridge, MA: MIT Press.

Quigley P (1999) Halo becomes albatross for Virgin. *Wireless Week*, November 22.

Resnick M (1994) *Turtles, termites, and traffic jams: explorations in massively parallel microworlds*. Cambridge, MA: MIT Press.

Silberman S (1999) Just say Nokia. *Wired*, September.

Smith B (2000) Welcome to the wireless internet. *Wireless Week*, January 31.

Starlogo homepage, MIT Media Lab. http://starlogo.www.media.mit.edu/people/starlogo/

Witzgall B and Kaye J (2001) Enhancing conversation through context output. *Projections: The Student Journal of Planning* at MIT. No. 2.

Part 3
From Ethnography to Use

Chapter 6
Seeing the "Rules": Preliminary Observations of Action, Interaction and Mobile Phone Use

Ged M Murtagh, Research Fellow, Digital World Research Centre, University of Surrey, England

One wants to make a distinction between "having the floor" in the sense of being a speaker while others are hearers, and "having the floor" in the sense of being a speaker while others are doing whatever they please (Harvey Sacks[1])

6.1 Introduction

The purpose of this chapter is to outline a discussion of the first phase of an ethnographic study of mobile phone use on train carriages. The chapter constitutes a brief exploration of some ordinary features of action and interaction as constitutive features of social action and technology use. As such the discussion is designed to be suggestive of how these initial findings can be analysed and developed further. The analytical orientation is informed by an ethnomethodological approach insofar as it is concerned with the rules of mobile phone use "as instructions for seeing" (Harper and Hughes, 1993) and understanding the "character of interpersonal communication" (Heath and Luff, 1993) in public places.

The chapter will outline some of the responses observed when people used their mobile phones on trains and suggest a way in which these behavioural phenomena might be analysed, described and understood. The discussion offers tentative suggestions on the basis of an, as yet, limited observational corpus. The overall aim of the discussion is to suggest a way in which observational investigations of mobile phone use in public, might be developed. The last section of the discussion provides some direction for this, with a consideration of how this kind of research might inform the design process of mobile technologies.

6.2 Analytical Background

The research is informed by investigations of non-vocal aspects of phone use. There are two reasons for this. First, as mentioned, the investigations are based on observational work only. Secondly (and perhaps more importantly from an analytical point of view) the focus on bodily gesture and eye contact emerged from the data when it was noticed that activities with and responses to mobile phone use were almost invariably non-vocal in nature. Examples from the research typically involved participants modifying their bodily position and direction of gaze when using mobile phones. The analytical task then lay in describing the intersubjectively problematic issue of non-vocal activities in conjunction with mobile phone use on train carriages.

Using a mobile phone involves the user engaging in a verbal exchange with another. Yet, so much of mobile phone use in public is organised through non-verbal action and interaction. It is suggested that these non-verbal aspects of phone use display the "unwritten rules" of usage behaviour in public. The following discussion will attempt to describe situations of phone use where "members' actions are features of the events their actions are accomplishing" (Garfinkel, 1967: 6).

Therefore, the focus is on some of the common sense considerations of social structures available to members to produce, sanction and maintain the orderliness of their everyday conduct. The purpose of this chapter is to go beyond focusing on how mobile phone behaviour "reveals an order of interaction" (Garfinkel, 1967), to identify some considerations available to members "that presuppose an understanding of that order" (*ibid.*).

It is suggested that an ethnomethodological approach is well suited to an investigation of this kind. Its analytical foundations are sensitive to the ordinary features of everyday action, the mundane, the seen but unnoticed aspects of daily life. Adopting this approach to investigate what is now common place activity (mobile phone use in public) shifts the focus toward the salient features of technology use when immersed in a situated order of social action.

6.3 Ethnomethodology, Technology and Situated Practice

In relation to a study of technology, an ethnomethodological analysis demands that the analyst attend to situated social practices, methodical procedures and accountable features of society's member's use of technology. In an investigation of the "social response" to modern technology (in this instance, mobile phone use), ethnomethodology places emphasis on the actual situated use of technology as an observable, visible and accountable feature of social action.

> They [ethnomethodologists][2] are thus concerned with issues such as the situatedness of work practices, the local deployment of knowledge, the assemblage of context, interactional contingencies, praxis, and the sense that these matters have in the accountable occasions of their investigation (Button, 1993: 1).

For the purposes of this discussion it is the themes of situatedness, local knowledge, context and interactional contingencies that feature significantly in the research findings. More importantly, the discussion is concerned to demonstrate how these aspects of social life are made available and visible to others in the ongoing process of social order. What is of particular interest is how these factors are sustained when a potential intrusion to that same social order is introduced, i.e. mobile phone use. Part of this concerns the relationship between rules, social action and technology use.

6.4 Rules, Social Action and Technology Use: The Case of Air Traffic Control

An exemplar of the approach taken here is found in Harper and Hughes' (1993) study of technology use in air traffic control. Although this study is an investigation of work practice, it provides a clear account of the utility provided by an ethnomethodological analysis of technology use and situated practice. As an exemplar account they focus on the question of the relationship between rules, social action and technology use. Part of their study concerns the situated accomplishment of rules as integral to the setting or context in which the action occurs. This is a thematic that is particularly relevant to the present investigation. Harper and Hughes suggest that the view that rules exist as external factors that govern or determine social action:

> ... fails to acknowledge that rules have to be applied within a setting such that what a rule or procedure means, what action falls under it, is a matter that has to be decided, judged, determined on occasions of its application. Social actors, that is, have to make judgements as to whether, *this* rule applies *here* and *now* in respect of *these* circumstances. (Harper and Hughes, 1993: 128)

In relation to the work practice of air traffic controllers and the technology they use, Harper and Hughes stress the importance of capturing the informal logic of the air traffic controllers as a constitutive feature of their working practice. More importantly they emphasise the interpretative practices, skills and competencies employed by the controllers to manage the contingencies and "unfolding details" of air traffic control. In this respect, they argue, the "rules" for the technology that controllers use is always contingent upon the interpretation of those same "unfolding details" of each situation. They illustrate some of the common sense considerations of social structures available to the controllers to produce, sanction and maintain the orderliness of their everyday conduct:

> What each source of information means is reflexively determined by the sense made of the others: a mutuality that is premised upon "learnt through experience" knowledge of controlling ... in short, tacit features of controlling activities. (Harper and Hughes, 1993: 138)

Harper and Hughes point out that ethnomethodology can offer an alternative (insofar as it is non-reductionist) insight into the interrelationship between user, environment and technology. They stress the importance of attention to actual situated practice, common sense understandings and informal logic in use to understanding working practice and technology use, as a potentially useful design tool.

This chapter borrows from part of the thematic of this study in that it is concerned with how rules (in this case rules of mobile phone use) are contingent upon the process of interpretation. The focus is on the situated accomplishment of rules as integral to the setting or context in which the action occurs. In this respect, determining a rule and the action that falls under it, is a matter that has to be decided, judged and determined on occasions of its application. As Harper and Hughes point out, the technology in air traffic control was used in conjunction with the social organisation of working activities. Similarly, when used in conjunction with everyday activities, mobile phones enter an already existing socially sanctioned public order. It is hoped that by briefly examining this issue and how ordinary members accommodate the advances of mobile technology in everyday life, some direction may be provided for the future design of mobile devices.

6.5 Displaying the "Rules"

The research question concerns the ambiguity surrounding the rules of mobile technology use in public places. The ambiguity mentioned concerns the problematic of defining the parameters of appropriate phone use within the setting of the train carriage. It is suggested that the parameters of appropriate use can be gauged through observing some non-vocal activities and responses to phone use. That is the case insofar as these responses are available to ordinary members to determine, assess, judge, attest and proclaim the meaning of an action with a mobile phone. Moreover, it is suggested that these same non-vocal responses provided ordinary members with resources to display the "rules" of appropriate mobile phone use. The responses discussed are illustrative of some general patterns observed; however they are derived from the first initial phase of observational study. Further investigation is demanded but it is hoped that this discussion may play some part in facilitating that.

6.6 Non-vocal Responses

From the start of the research investigations the interactional significance of eye contact (where mobile phones were used in close proximity with others) became readily apparent. Whether it be managing the contingencies of a busy pedestrian flow or communicating whilst in transit, eye contact emerged as one of the key non-vocal responses that demanded further investigation. Many of the recurrent patterns of non-vocal interaction are described in the study of interaction, eye contact, bodily movement and their relationship to one another. This area of study has provided for a considerably extensive literature in sociology[3] and social psychology.[4]

In most instances of mobile phone use on the train, eye contact with others co-present was avoided. Indeed the occasions where people would make eye contact with others whilst using the phone, were comparatively few. This may have something to do with the use of eye contact in public places as described by Goffman (1963). Goffman identifies what he refers to as "civil inattention" as one of the rules that govern our behaviour in public places in the ongoing concern with the "observance of social propriety" (Burns, 1992). In public settings, Goffman argues, a state of mutual gaze is avoided. And so it was with phone use – mutual gaze between phone user and co-present others was generally avoided.

However, two kinds of behavioural response were noticed when somebody's phone rang.[5] Firstly, there were a number of instances of those receiving the call averting their eyes away from the direction of others toward a neutral space. They would maintain this direction of gaze throughout the duration of the call, occasionally glancing elsewhere. For example, some (on receipt of a call) could be seen to initiate a downward head movement whilst simultaneously raising the phone to the ear on answering. As the conversation proceeded they would maintain a downward directional gaze, whilst others would raise the head slowly glancing in all directions, although eye contact with others was fleeting.

Consistent with Goffman's line of reasoning one might begin to look at these responses as a means to manage the potential embarrassment surrounding the public audibility of a private conversation. Receiving a call in a setting where others are present may be viewed as an infraction of the rules of public settings where normally there is a preference for civil inattention. Subsequent investigations into eye contact and interaction have formulated rules of relevance. Goodwin (1981), for example, has suggested that the direction of gaze is a means through which members can display the statuses of hearer or speaker. Gaze, he suggests, is sequentially related to verbal utterance. For example a speaker will often seek the gaze of the recipient prior to a verbal utterance. Moreover gaze is often an indication of interactional availability vis-à-vis speaker or hearer status.

On several occasions those co-present would glance at the phone user and then return to their activities or separate lines of concern. Goffman's notion of "civil inattention" is useful when analysing this kind of response. Typically, in public settings people collaboratively display a disengagement from each other's activities and behaviour. In many instances this response concerns the matter of separating the public from the private. One could adopt Goodwin's line of analysis to suggest that both parties to phone use within close proximity mutually accomplish the relevance of that activity in a "seen but unnoticed" fashion. The ringing of the phone draws attention to the phone user; however the public/private boundary of that setting is organised through eye contact. More importantly the direction of the gaze is available to the participants as a means of displaying and confirming the "rules" of verbal exchange using a mobile phone in public.

6.7 Body Movements

It is worth noting that posture or bodily gesture also appeared to be a significant factor in relation to phone use. Like changing the direction of gaze, the turning of the head and upper body away from others co-present was also a feature of mobile phone use in the train. It is suggested that these behavioural responses are indicative of, or display, not only the moral implicativeness of speaker and hearer status but also the subtle complexities of interaction and its embeddedness within particular social contexts.

Heath and Luff (1992) suggest that eye contact and bodily movement can be used as an interactional resource to organise the activities of co-present others. Co-interactants, they argue, assume a reciprocity of perspectives "to assess the impact of their own conduct, and to design action with respect to the way in which it is likely to be seen and treated" (Heath and Luff, 1992: 315–346).

Thus head movement is used in conjunction with eye movement and the two could be seen as related. So far, the discussion has qualified the purpose of these gestures as effecting some sense of privacy in public settings whilst using a mobile phone. The argument rests on three premises:

- The turning and looking at another is interactionally significant.
- Turning and looking away from co-present others suggests that one is otherwise engaged.
- The activity of mobile phone use is essentially a matter of managing privacy where others are co-present, and the mobile phone user is engaged in displaying deference to others.

However it doesn't necessarily follow that looking away from someone indicates a discreet disengagement from what they are saying or doing, nor that mobile phone use in public is essentially a matter of managing privacy. It may be a matter of managing privacy; however, the question is, when does it become a matter of privacy and how is the situation constituted as such. Much of this depends on how others co-present respond and the analytical weight that can be attached to the interpretation of these responses.

6.8 The Unanswered Phone

The potential intrusiveness of the phone begins the moment that it starts to ring. Indeed, the ringing is sometimes attended to by others through a glance in the direction of the ringing. However, a number of instances were observed where the ringing of the phone continued for a longer period. In these instances there were two noticeable responses. The first is that those co-present would begin to stare in the direction of the ringing. The second is that others co-present would check to see if it was their phone that was ringing. Although still tentative at this stage, these observations were made on several occasions and they seem to suggest that there is a normal expectation or shared common understanding with regard to the length of time taken to answer the phone where others are present.

There are a variety of ways in which this might be analysed. Schegloff (1986) has suggested that the ringing of the phone can be seen as the summons part of a summons/answer sequence. He makes this observation from his analysis of ordinary conversation (with Harvey Sacks) as a sequentially organised phenomenon. This can be evidenced through observing one of the foundational building blocks of social interaction, i.e. the adjacent pairing of conversational utterances. Common examples include question/answer sequences, invitation/acceptance-refusal, request/rejection.

A given sequence will thus be composed of an utterance that is a first pair part produced by one speaker directly followed by the production by a different speaker of an utterance which is (a) a second pair part, and (b) is from the same pair type as the first utterance in the sequence (Schegloff and Sacks, 1974).

Thus ordinarily, answers follow questions, greetings are reciprocated and invitations or requests are accepted or rejected. In this way adjacently paired utterances act as an interactional form of social control and serve to solicit an expected response from another. This is not to say that speakers and hearers are constrained by the sequential implicativeness of adjacently paired utterances (e.g. greetings are not always reciprocated, questions are not always answered) but that generally both speakers and hearers orient to the assumption that adjacent utterances are to be heard as related.

> The class of cases, together with related instances in which second speakers' responses are treated as "not answering the question", demonstrates that questioners attend to the fact that their questions are framed within normative expectations which have sequential implications in obliging selected next speakers to perform a restricted form of action in next turn, namely, at least to respond to the question with some form of action (Heritage, 1984).

These sequentially implicative expectations act as a device to solicit a next action. An answer is "conditionally relevant" to a prior question, a question establishes the conditional relevance of an answer. The absence of an answer to a question is a noticeable action. To render the absence as morally accountable the summons might be repeated until the answer is obtained (Goodwin, 1981).

In the same way, a phone that goes unanswered is, by analogy, a morally accountable phenomenon evidenced by the facial gestures and the direction of gaze of others co-present. Like a question, the ringing of the phone is designed to solicit a response from the other. Where that response is not forthcoming it is made accountable both by the caller (who awaits the answer) and also, in this instance, by those co-present who display, through gaze and bodily movement, the expectation to answer somebody's call.

On this view it is suggested that the unanswered mobile phone is indicative of some aspects of the social organisation of (what used to be a) private activity within the public realm. The ringing phone has sequential implications for the receiver to perform a next action. The absence of that next action is an infraction of the "conditional relevance" of the initial summons through the ringing phone.

6.9 Other Common Sense Considerations

In a later article Schegloff (1986) discusses in more detail the kinds of inferences available to co-participants where a phone goes unanswered. He points out some of the interactional issues where persons other than the caller and the person being called are present. For those others, multiple rings are a source of inferential topicalisation as are very few rings.

But this under-specifies the relevant organisation, because which of several persons present will be involved and how, is also orderly, and requires various kinds of analyses by the persons in such an environment to find what is sequentially relevant for them (*ibid.*).

The fact that the phone remains unanswered in public makes that absence inferentially available to others. The mobile phone is a device that can be switched on or off. Thus staring in the direction of the unanswered phone may have less to do with the understanding of the conditional relevance of an answer to a summons. Rather it may have something to do with the mundane fact that someone has not displayed the courtesy of switching their phone off in public where others are in close proximity. In this way mobile phone use is re-constituted as a matter of social etiquette. Ultimately, however, the unanswered may generate no response whatsoever.

6.10 Time

Another feature of mobile phone use that is inferentially available to others is time. Time may be a consideration in relation to the call itself. In one particular instance on a train travelling at peak time a man, A, sat opposite a person on their mobile phone, B. A began to display his discomfort after B was on the phone for approximately 20 minutes. The length of time taken on the phone provides a "legitimate" resource for non-vocal complaint work in suggesting a normal expectation regarding the length of time on the phone where others are in close proximity. It is worth noting that this display was immediately available as an expression of discomfort. Precisely what that display of discomfort related to, is open to debate. Suffice to say the length of the call may be implicated in A's assessment of the situation. It is certainly available as a resource to judge, assess and determine the appropriateness of the behaviour.

The issue of acceptability of phone use may be related to time in another way. Usage at peak times may be considered more acceptable due to the nature of work-related demands and the current implementation of mobile devices and work-related tasks at these times. The element of time certainly provides for a resource available to members to determine the parameters of usage behaviour. In other words it might be assumed that for members, calls at peak times are on the whole work-related. By contrast non-work-related calls in work-related hours may be viewed very differently. Much of this however hinges on the conversational topic that ensues.

6.11 Conversational Topic

The aforementioned example of A and B was of a call that took place in peak hours travelling time. As mentioned, the display of discomfort came after considerable time but also coincided with a change of conversational topic. The person using the phone had, up until that point, made work-related calls. He then made a social call where he was talking to a female friend about his activities at the weekend and proceeded to converse with her about her time in Italy.

Thus conversational topic is also a resource available to members to constitute a response to mobile phone use. In this instance it might be inferred that the topic of the call was considered inappropriate within the setting or indeed at that time. If one may be permitted a gloss, these factors (and others) can be seen as resources with which members can determine the "legitimacy" of an interruption of their time in transit. The non-vocal responses discussed are a means through which members can display the appropriateness and legitimacy of phone activity. Moreover, they are also a means through which members can invoke the "rules" of phone use as a public display.

As with the Harper and Hughes study, the emphasis in this brief discussion has been placed on the interpretative practices, skills and competencies employed by the phone users and co-present others to manage the contingencies and "unfolding details" of phone use on train carriages. Rules and their meaning, it has been argued, are to be decided, judged and determined by members on occasions of their application. That is to say users and non-users make judgements as to whether, "*this* rule applies *here* and *now* in respect of *these* circumstances" (Harper and Hughes, 1993: 128).

6.12 Future Research and Future Technology: The Issue of Design

This piece of ethnographic fieldwork has attempted to gauge the "situatedly specific activities" (Harper and Hughes, 1993: 142) of mobile phone use as "seen from within" (*ibid.*). Insofar as it addresses a technological device that is continually modified and innovated, it is worth pausing to consider some design implications. That is not to say that there is a direct line of sight from a piece of ethnomethodologically informed ethnographic research to design and innovation. The relationship between ethnography, ethnomethodology and design has been dealt with in more depth elsewhere.[6]

Rather, the purpose of this section is to make some general comments (on the basis of the fieldwork) that might be considered in relation to proposed future developments. From the outset it goes without saying that further investigations are required to enhance our understanding of non-vocal activities in relation to the social "rules" of phone use. Moreover, this necessitates further discussion of the implications of non-vocal activities for emerging technologies.

As mentioned, whether it be managing the contingencies of a busy pedestrian flow or communicating whilst in transit, eye contact emerged as one of the key non-vocal responses. Moreover, this non-vocal response was observed to be an important means through which members organised, structured and managed their use of and response to mobile technologies with co-proximate others on train carriages. Thus, the precise co-ordination of gaze and bodily and spatial orientation has important implications for further applications of interactive technologies.

For example, accounting for the precise co-ordination of gaze and bodily and spatial orientation in public places may provide useful insight for the design and modification of screen size in relation to internet services or the mobile video phone. Moreover investigations might be carried out on the impact of interactive mobile devices as a potential disruption to patterns of non-vocal interaction in public settings.

The setting itself, its spatial and structural arrangements, was also a key factor in relation to mobile phone use. On a number of occasions phone users were observed moving to corridors between carriages or the space that joins the carriages. In one instance a phone user moved to the space between the carriages six times to make and receive calls. Presumably the user did not want their conversation to be audible to co-present others. However, the reverse was the effect as the user could be heard halfway down the carriage where the researcher was sat. The reason for this is that the ambient noise is greater in the space between the carriages. Thus the phone user had to raise their voice so that the person they were talking to could hear them.

It is hoped that the kind of research discussed here might inform design in terms of the adaptive and responsive features of mobile phone use in public. An ethnomethodological perspective can be employed to deliver a reappraisal of the relationship between system and user. That is to say, in analysing some of the issues discussed, it is hoped that an ethnomethodological analysis may provide "a better 'goodness of fit' between the technology and the working environment in which it is located" (Harper and Hughes, 1993: 143).

6.13 Conclusion

This chapter has attempted to articulate some of the ways through which non-vocal complaint work can be "legitimately" displayed in response to mobile phone use in public. The focus has been on the common sense reasoning procedures available to members to render behaviour with mobile phones accountable.

This order is produced and reproduced by ordinary members of society, in and through the subtle but significant aspects of social interaction, as they conduct the practical affairs of their everyday lives. The point is that through the introduction of the mobile phone, phone use behaviour is made available to others thus providing a whole host of inferences to treat particular phone behaviours as accountable, acceptable, unacceptable, intrusive, etc. Identifying these matters within a specific context is dependent upon members' common sense knowledge of the peculiar features of the mobile phone itself and their knowledge of the settings within which usage takes place.

The responses to mobile phone use may include some of the features discussed above. They may include all of them, they may include none of them. Much of this hinges on what is considered relevant to those present at a particular time, in a particular setting, etc. The observational data alone only provides so much analytical weight. Nevertheless, it is hoped that this chapter addresses a foundational concern for future research in a respecification of the relationship between environment and user through the study of actual situated use.

Notes

1. Cited in Charles Goodwin (1981) *Conversational organisation: interaction between speakers and hearers*. New York: Academic Press, p. 59.
2. My insertion.
3. The foundational sociological contribution comes from Erving Goffman's work. See for example his *Behaviour in public places*, Free Press, New York (1963). See also the work of Charles Goodwin (1981) *Conversational organisation: interaction between speakers and hearers*, Academic Press, New York. See also Christian Heath (1986) *Body movement and speech in medical interaction*, Cambridge University Press, Cambridge. George Psathas (1990) "The organisation of talk, gaze and activity in a medical interview" in George Psathas (ed.), *Interaction competence*, University Press of America, Maryland.
4. Within the psychology literature see Adam Kendon, in particular 'Some functions of gaze direction in social interaction', *Acta Psychologica* 26 (1967), 22–63 and Kendon (1990) *Conducting interaction: studies in the behaviour of social interaction*, Cambridge University Press, Cambridge.
5. My intention is not to privilege these responses over any others but to point to the fact that throughout the observational period they appeared to be the most common.
6. See for example, Button G and Dourish P (1998) On "Technomethodology": foundational relationships between ethnomethodology and system design. *Human Computer Interaction* Vol. 13, 395–432.

References

Burns T (1992) *Erving Goffman*. London: Routledge.

Button G (1993) *Technology in working order: studies of work, interaction and technology*. London: Routledge.

Garfinkel H (1967) *Studies in ethnomethodolgy*. Englewood Cliffs, NJ: Prentice Hall.

Goffman E (1963) *Behaviour in Public Places: Notes on the Social Organisation of Gatherings*: New York, Free Press.

Goodwin C (1987) Unilateral departure. In G Button and J Lee (eds). *Talk and social organisation*. Clevedon: Multilingual Matters.

Goodwin C (1981) *Conversational organisation: interaction between speakers and hearers*. New York: Academic Press.

Harper RHR and Hughes JA (1993) What a f-ing system! Send 'Em all to the same place and then expect us to stop 'Em hitting. In G Button, *Technology in working order: studies of work, interaction and technology*. London: Routledge.

Heath C (1986) *Body movement and speech in medical interaction*. Cambridge: Cambridge University Press.

Heath C and Luff P (1992) Media space and communicative assymetries: preliminary observations of video mediated interaction. *Human Computer Interaction*, 7, pp. 315–346.

Heath C and Luff P (1993) Disembodied conduct: interactional asymmetries in video-mediated communication. In G Button, *Technology in working order: studies of work, interaction and technology*. London: Routledge.

Heritage J (1984) *Garfinkel and Ethnomethodology*. Cambridge: Polity.

Kendon A (1990) *Conducting interaction: studies in the behaviour of social interaction*. Cambridge: Cambridge University Press.

Psathas G (1990) The organisation of talk, gaze and activity in a medical interview. In G Psathas (ed.) *Interaction competence*. Maryland: University Press of America.

Schegloff EA (1986) The routine as achievement. In *Human Studies*, 9, pp. 111–151.

Schegloff EA and Sacks H (1974) Opening up closings. In R Turner (ed.) *Ethnomethodology*. Harmondsworth: Penguin.

Schegloff EA (1979) Identification and recognition in telephone conversation openings. In G Psathas (ed.) *Everyday language*. New York: Irvington.

Chapter 7
Local Use and Sharing of Mobile Phones

Alexandra Weilenmann and Catrine Larsson, Viktoria Institute, Göteborg, Sweden

Here's an object introduced into the world 75 years ago. And it's a technical thing which has a variety of aspects to it. It works only with voices, and because of economic considerations people share it, so that there are not yet things where you can call up a particular person and get them, or get nothing. Now what happens is, like any other natural object, a culture secretes itself onto it in its well-shaped ways. It turns this technical apparatus which allows for conversation, into something in which the ways that conversation works are more or less brought to bear. So, there evolves from the introduction of the telephone, a collection of rules about its use ... Sacks (1992/1995: 548)

7.1 Introduction

As researchers interested in the use of mobile phones, everyday activities provide us with a rich resource to observe phone usage. We often see people using their phones on the bus, in parks, cafés, on bikes, in the streets, waiting for the cash machine, in shops – nearly everywhere.[1] This public use of the mobile phone provides us with an invaluable resource for looking at the everyday, actual use of this technology.

As this book shows, the social and interactional aspects of mobile phone use have only recently become a topic of interest. Along with the chapters in this book, there is work looking at how mobile phones affect the urban society (Kopomaa, 2000), how families and their teenagers co-ordinate their lives together using mobile phones and how the ownership and payment structure works within these families (Ling, 1999a, b; Ling and Yttri, in press), the use of mobile phones in relation to the other media (Koskinen, 1999; Coogan and Kangas, 2000), and how mobile phones are used by mobile professionals as one of several available resources for mobile work (Wiberg and Ljungberg, 2000).

These studies all contribute to understanding the use of the mobile phone. In this chapter, however, we focus on a relatively neglected aspect of mobile phone use. That is, the *local interaction* of mobile phones, the ways in which phones are used and shared in the local situation of use. From field studies of public use of mobile phones among teenagers in Sweden, we report on how the mobile phone has come to be used as a tool for local social interaction, rather than merely as a device for communication with dislocated others. The collaborative nature of mobile phone use is very evident from our observations. Mobile phones are often *shared*, in various ways and for various purposes. We examine how this sharing is accomplished. Finally, we discuss how this empirical field data can be of use when designing new mobile technology and services for young people.

7.2 Mobile Phone Research

Mobile phones are often used in public spaces, and many of the observations we collected for this chapter come from urban situations. Kopomaa's (2000) book *The City in Your Pocket* is of particular interest in the way it describes the impact of this "new everyday appliance" on urban life. He describes how city social life has changed when public mobile phone use develops. Kopomaa gets most of his data from interviews, although these are supplemented with photos taken in public of people observed talking on the phone. Kopomaa's book, therefore, focuses almost exclusively on using mobile phones for "phoning", i.e. making and or receiving phone calls, and the book neglects (although it does reference) other phone practices. For instance, he does not report having made any observations on SMS messaging, the practices of which he discusses in the book.

More specifically with regard to teenagers, this segment has received particular attention in the Scandinavian countries. For instance, Koskinen's (1999) work is concerned with teenagers' use of SMS and other "asynchronous messaging with mobile systems" as well as with the convergence between mobile phones and other media. Research about Finnish teenagers' use of mobile phones and the internet is studied in the NUFIX project (Coogan and Kangas, 2000). In Norway, Ling and his colleagues have done many interviews and surveys with teenagers and their families to get the picture of use, ownership and conception of mobile telephony (Ling, 1998, 1999a, b; Ling and Yttri, in press). There is also a Swedish study of teenage behaviour focusing on identifying mobile internet services for young females.[2] These studies all show the great immersion of mobile phones among teenagers, and point to the importance of this device in the life of this age group. Of course, Scandinavian adoption of phone use is behind Japan where mobile phones are commonly used for many purposes other than calling. The mobile phone operators offer a large number of features and services. The introduction of wireless internet access, and especially the NTT DoCoMo's I-mode services, has made it possible to use mobile phones for sending and checking emails, chatting, etc. Teenagers in Japan are said to be the driving force in much of the use of mobile phones (Mitsuoka *et al.*, 2001). However, unlike Scandinavia there is little research (at least in English) exploring the social aspects of these new devices. It is likely that many other countries are approaching what Scandinavia and Japan are experiencing. It has been estimated that mobile devices will outnumber televisions and personal computers within a few years. The mobile phone is already changing from being a tool for voice communication into being a more general communication device. Whatever these devices will look like in the future, the mobile phone in some form is here to stay.

7.3 In the Field

This chapter is based on fieldwork performed in central Göteborg, Sweden. We have focused on inner city public use of mobile phones among teenagers. So far there are no comparisons made with teenagers' use of mobile phones in other cities, regions or countries. The fieldwork has been carried out in a wide range of places and situations, such as cafés, public transport (bus, tram, train), an amusement park, shopping malls, etc. In all these places, teenagers can often be seen passing time or "hanging out". All these places are public. The reason behind focusing on public use was originally methodological: we wanted to be taken for ordinary individuals and not as researchers focusing on naturally occurring interaction. This would not have been possible (or at least would have been very difficult) had we chosen to study teenagers' use of mobile phones in a private setting.

Being an anonymous observer in a public space has interesting methodological implications. When looking at the local social impact of telephone use, the researcher only has access to the same information that the other participants have. For instance, if we are sitting on a bus and someone gets a phone call, we are in a similar position as other people on the bus in reacting to the phone call. The ongoing interaction is as observable to us as it is to any other person present. This also means that we can only gather data that the members themselves actually make available; we felt that was important for ethical reasons. If someone talking on the phone did not want anyone in the surroundings to hear, we assumed they would simply talk quieter, making it impossible for us to hear, or choose not to have the conversation at all.[3] Our field study work was designed to be open ended. We generally wanted to address the question: "What do teenagers do with mobile telephones?" We did not want to make too many prior assumptions about the use, and by that miss out on what was really going on. Indeed, as we will discuss later in this chapter, by using this naturalistic technique we uncovered data on a phenomenon so far neglected in the literature – the local sharing of mobile phones.

When doing fieldwork on public transport, we would generally sit on a bus or tram and observe the use and handling of phones. In cafés, we would sit at a café table like other café visitors. In the other environments, the amusement park, the shops, in town, we tried to do what people in these places generally were doing.[4] We did fieldwork during different times of the day and week, which means that the data encompasses late Saturday night as well as early Monday morning, and everything in between.

The observations were documented in field notes. In these notes, we made detailed descriptions of all observable events including teenagers handling their mobile phones in any way. For most of the observations, we worked as a team of two observers in the field at the same time. This had several benefits. First, it made it possible to observe more *details* of the ongoing interaction. Since we had to collect all our material in notes rather than recordings,[5] two persons' observations could give a more detailed picture. Further, when analysing the data, it was an advantage that we both had observed the instances, since the only material was documented in notes. We believe that this gave more a true recollection of the situation, and thus a more true analysis.[6]

The field observations were then analysed, looking for certain themes in the data, themes that seemed to have relevance for the teenagers themselves, rather than just for us as researchers. This chapter focuses on how teenagers do things together with their mobile phones, which means that many of our observations of solitary use are not discussed here. In the presentation of our observations the teenagers are named A, B, C, etc., in the order they appear in the description. The mobile phones are numbered 1, 2, 3, etc. in the same manner. Approximate times of the day are given, as well as a general description of the place in which the observation was made. The translations from Swedish have been made by the authors.

7.4 Sharing Mobile Phones

The mobile phone is often described as a personal phone. In research as well as in the design discourse, there seems to be an underlying assumption that mobile phones are used by individuals for remote communication purposes. In this chapter, we hope to show how the mobile phone has become something more; it has become a tool for collaborative interaction in the local environment. Among the teenagers we have studied, the phones are not just treated as personal, and the calls and other communication are not treated as private. Rather, there is much work going on to render the communication "public", enabling several people to take part in it. The remote communication, i.e. the phone calls they receive or make, as well as the SMS messages they receive or send, are accounted for in the ongoing local interaction. Teenagers thus share the communication they take part in with their co-present friends. Not only the communication but also the phone itself is often shared. These findings question the notion of the mobile phone as merely a personal phone used for remote communication. In this section we examine the ways in which phones are shared within the local environment.

We will begin by looking at *minimal forms of sharing*; how the content or information on the phone is made accessible to others in various ways, without sharing the phone itself. In the remainder of the chapter then, we will deal with a more *"hands-on" sharing*, where the phone itself is shared and handled by more than one person.

7.4.1 Minimal Forms of Sharing

In this section, we examine examples of a minimal form of sharing of mobile phones. What is shared in these instances is only information on the phone, in this case SMS messages. There is no physical form of sharing of the phones, i.e. the phone remains in the hands of one person. The two strategies for minimal forms of sharing SMS messages presented are sharing by reading the message aloud, and sharing by showing the display to others. Both strategies are ways to let friends (and perhaps others in the surroundings) take part of or part in personal communication – ways to render private information displayed on a very small screen accessible to others. These strategies seem to occur when teenagers try to engage others in the SMS messaging activities they currently are involved in.

In the first excerpt from the field notes, a girl is sharing her SMS message with her friends, while writing it. She first shows the display to her friends, and then reads aloud from her message. This one girl is responsible for all the physical interaction with the phone, but how she relates to her friends while using the phone is important for understanding what is going on:

> *Excerpt 1: Tram, evening*
> Three girls are sitting on a tram. One girl, A, is writing an SMS message. A turns to B, who is sitting next to her, gives B a light nudge, and says "hey". She shows the display to B. A deletes a few letters, and then continues to write the message. She says with a whiny voice:
> A: I don't wanna send this.
> A then begins to read aloud parts of her message:
> A: "I want to have a home party. I'm leaving soon you know."
> Presumably she now sends the message. She then puts her phone in her purse. Shortly after this, her phone rings. She exclaims "No". She picks up her phone, and without looking on the display, gives the phone to her friend, B, and says:
> A: please, can you get it?
> B pushes the phone away, refusing to answer it. A answers the call. A talks to someone about the home party. She ends the phone call after a short conversation.
> A: [to the others] I hate (him)! Shit! [sighs]
> B: What did he say?
> A: Nothing! [she turns toward the window, crosses her arms and sighs]

This case basically has two parts: the first part is where girl A is writing her SMS message, and the second part begins when the phone rings in her purse.

In the first part, the girl is trying to involve her two[7] friends in the production of an SMS message. What could be a solitary activity, writing a text message displayed on a very small screen, becomes a group activity. A shares her message by first actually showing the text (or the part of the text currently visible on the screen) to her friend B. This is done quite subtly, by a light nudge with the elbow and displaying the screen, thus indicating she wants B to read it. In this case, B does not seem to be very interested in the message; this is actually a point where B could take A's phone in her hand, to be able to read the whole message and involve herself in its production. For some reason, B does not do so. Shortly after this, A begins another strategy to let/make others know about the content of the phone; she reads aloud from her message. This could be because she wants more involvement from the others, and could not get it by simply showing the message. Reading aloud from the screen is thus a way to render the message on her screen "publicly" accessible (Heath and Luff, 2000:67). In the second part of this excerpt, A's phone rings shortly after she has sent the message. A now wants to involve her friend B in this phone call; A asks B to answer the phone, but B does not accept. A answers herself. After the conversation on the phone, she comments on the call to the others. Talking about the content of the call or commenting on the caller, is very often observed among teenagers.

In the next excerpt, we see another short example of the strategy of sharing content by showing the display to others:

Excerpt 2: Café
A girl comes back to her table after (presumably) having gone to the rest rooms.
She is holding a blue phone in her hand. She gives B a nudge with her elbow, shows
her the display, and then puts the phone on the table.

From mere observations it is impossible to say whether the content A is showing
consists of an SMS message or some other information on the phone. Regardless of
content, the strategy is the same as in the example above where the content shared
was an SMS message under production. In this case also the strategy for letting
others take part of the message is to show the screen, and this is done by the elbow
nudge (as in excerpt 1). The phone is then put on the café table, maybe as a way to
demonstrate that further interaction with the phone is possible from B's part.

In these two excerpts we have seen two different strategies for minimal forms
of sharing content on the phone. The first was reading aloud from the message, and
the second to show the display with the message. Both strategies are ways to try
to involve co-present friends in the remote communication one is engaged in.
While the content of the phones has been shared, the phones themselves have not.

7.4.2 Taking Turns

The two examples discussed so far both involved one person having a phone and
sharing the content of it in various ways. However, this was a minimal form of
sharing; there was no physical sharing or exchange of the phones themselves. In
the next excerpt from our field observations, the sharing of content is done
through several people actually handling the phone. This is a different form of
sharing which is more "hands-on".

In the first sequence, several teenagers share one phone call through taking
turns holding the phone and talking to the caller.

Excerpt 3: Saturday evening, tram
Four girls, about 13–14 years old, are sitting on the tram. They are sitting two and
two, facing each other. A phone that girl A is holding rings and she answers. She
says: "Yes, hold on", and gives the phone to B, who is sitting to her right. B talks for
a couple of minutes with the person on the other end, E. B says to the other girls,
while holding the phone a little bit from her head:
B: She says it's more fun at the other place.
C gets upset and reaches over to get the phone. C starts talking to E:
C: What do you mean by saying that? You shouldn't say that. Now everyone is
looking at the floor. You shouldn't say that.
After a short conversation, C ends the conversation by saying "bye, kiss, kiss". She
hands the phone over to B again, who continues talking to E.

This excerpt begins by A receiving a phone call that seems to be intended for
someone else, namely B. B thus takes the phone. After talking to the caller for a
while, B seems to want the others to be involved in the conversation; this is done
by sharing parts of the content of the conversations with the others ("she says it's
more fun at the other place"). This is similar to the strategy of reading aloud from

a message (cf. excerpt 1), in that it is a way to try to involve others in the remote communication. Apparently, from C's reaction, this is something that needs to be acted upon; C reaches over to get the phone. C then becomes the third person to be involved in this conversation by actually handling the phone itself.

From the observable actions described above, it is not possible to say who is the owner of the phone. It could be A, who is holding the phone when it rings. In this case it is noteworthy that the caller, E, knows whom to call when she wants to get hold of B. We have seen other such cases, where the caller by calling one person's phone, gets to talk to several people rather than just one. It could also be that the phone belongs to B; it is B that the caller gets to talk to. And it just happens that A at the moment is holding B's phone and therefore answers it when it rings.

In either of these two possible cases, the use of the phone shows us how the phone is shared within a group. We can see how several people take part in one phone conversation. This is an interesting type of multiparty talk, where one person (the caller) has limited access to the other participants in the conversation (those on the receiving end). It follows that the group on the receiving end only have access to the phone conversation through the person who is currently talking on the phone. What is notable is how little negotiation these young girls do in order to know how the phone should go around in the group. The phone moves from one hand into another in what could be called a *turn-taking system of mobile phones*. Using this analogy from conversation analysis (Sacks *et al.*, 1974), it is interesting to examine how and when others are brought into the conversation. How is the next user of the phone selected? What does C, in the above case, do in order to get the phone in her hand and begin talking to E? However, one suggestion could be that B, by holding the phone away from her head and talking to the others, makes it possible for someone else to self-select as taking the next turn using the phone. To pursue an analysis of this, there clearly has to be more detailed material, i.e. audio and video recordings. (We return to this point in the discussion section.)

Note also that girl C is trying to convey to E, the caller, the mood of the whole group. She does this by her metaphorical "looking at the floor". Whether they are really looking at the floor or not, whether they are sad or not, what is interesting is that C makes the whole group take part in the conversation. It is not only C that gets upset because of what E has suggested; it is the entire group.

From this example, it seems clear that the notion of the phone as merely a private resource for communication does not apply among teenagers. In this case the phone is shared to enable a whole group to talk to a remote person. Instead of one person talking and "shielding" her/himself from the group while doing it, everyone present involves themselves, and are allowed to involve themselves, in the ongoing conversation.

7.4.3 Borrowing and Lending of Phones

Several times when doing field observations, we had to reconsider our initial ideas about who the phones belonged to. The mobile phones passed through so many hands, that we had to stop and ask ourselves "whose line is this anyway?". Initially, we assumed the first person observed to be handling the phone to be the owner of the phone. It turned out to be a lot more complex. Many teenagers handle each other's phones, using them for various purposes, and it is sometimes impossible to understand from observations to whom the phone actually belongs. This borrowing and lending of phones seems natural to the teenagers themselves. These observations suggest that the mobile phone is a *collaborative resource* for teenagers, rather than just a personal phone. Below are a few examples of how teenagers borrow each other's phones for calling, answer their friends' phones when they ring, and even carry around each other's phones.

We begin with an example where one girl uses her friend's phone for calling although it later turns out that she has her own phone:

Excerpt 4: Café, Thursday evening
Two girls are sitting at a small café table, facing each other. A receives a phone call. She answers. She talks for a while before ending the conversation. She holds the phone in her hand, and shows the display to B. B takes the telephone and calls someone on it. She makes a very brief conversation. She gives the phone back to A. They light two cigarettes.
B now takes a phone out of her purse. This phone is ringing, playing the song *Come on baby light my fire*. After a short conversation she puts the phone back in her purse again.

A uses the same strategy here as described above for sharing of content – showing the display to her friend. This seems to prompt B to take the phone in her hand. Perhaps A was showing a phone number to B, so as to encourage her to call someone. In any case, B uses A's phone to make a phone call. We could not hear the conversation between the two girls in this case, thus we do not know if B received a verbal permission to use the phone. The phone call B makes is very brief. One reason for this might be because the phone call is costing A money and B therefore knows to be polite in keeping it short. The economic aspects of phone use are likely to be an important factor in the teenagers borrowing and lending phones.

Surprisingly though, in the above example, it turns out that B actually has a mobile phone of her own. This becomes evident when it rings. So why did B use A's phone to call and not her own? This could be because B was using a pay card on her phone and the card was empty, making it possible to use the phone for incoming calls only. It could also be simply because A's phone was at hand; it was more convenient to use that phone. However, this does not explain the economic decision to let B call from A's phone, unless it was that the phone call B made from A's phone was considered more related to A, and therefore she should pay for it.

In another similar example, one girl is using someone else's phone for calling although she has her own phone. What distinguishes these examples is that the girl above has quite a lot of control over the lending of her phone, since the person to whom she is lending her phone is sitting next to her, whereas in the following example this is different:

Excerpt 5: Tram, Friday at noon
A dozen high school teenagers get on the tram at a stop very near a high school. One girl, A, is holding a small pink telephone in her hand when she gets on the

tram. She has just finished talking on it and is now pushing the buttons and looking at it. When she has got on the tram she walks over to another of the girls, B, who has also just got on, and gives the phone to her. B says "Thank you", takes a quick look on the display and puts the pink phone in her bag.

A walks over to stand and talk with some other people. She now takes a green phone out of the front pocket of her trousers. She makes a phone call from this phone. After a short conversation she finishes, and puts the phone back in her pocket. She tells the others that she is going to meet a person later.

B has let A use her phone. When they enter the tram they are apart, which means that B cannot see what A is doing with the phone. In fact, B cannot be sure that A gets on the tram at all. This implies a trust in her friend. During the tram ride A and B are standing in different groups. When A has finished using B's phone she walks over to return it. It now turns out that A actually has a phone of her own, or at least another phone available to use: in her pocket. In this case, as in excerpt 4, we did not get access to the communication between the borrower and the lender of the phone prior to the sharing. In both of these cases, the girls use their own phone directly after borrowing a phone. This strikes the observer as odd, but apparently it is not treated as unusual among teenagers.

In the next example, two girls are involved in the sharing of phones. They are at a café, but not sitting together. The girl currently in possession of the phone has to take action when it rings:

Excerpt 6: Café, noon
A girl sitting at a table at the far end of the café has a phone that is ringing quite loudly. The girl, A, stands up and begins running across the café, while holding the ringing phone in her hand. She runs to the other end of the room, ten tables away. Four ring signals have now been heard. She gives the phone to a girl, B, sitting at a table. B answers the phone. A walks back to her table. B talks for a while, then puts on her coat and leaves the café together with her friend, and the phone. A remains seated in the café.

This girl's running with the phone ringing confused us at first, because we had seldom observed teenagers leaving the ongoing event to answer a phone call. When we saw her giving the phone to another girl at the café, we understood that she did not answer because she did not take this call to be intended for her.

In the excerpt below, three girls are interacting with two phones. Note how one of the girls demonstrates her thanks when receiving the phone:

Excerpt 7: Café, Wednesday afternoon
C holds a small red phone, 1, in her hand; she pushes the buttons and looks at it. She puts it on the table after a while.
B takes another phone, 2, from her purse, and gives it to A who hugs her. A makes a short phone call from phone 2. She gives it back to B who takes a quick look at the display.
B now gives phone 2 to C, who makes a phone call from it.
While C is talking, phone 1 rings. A quickly answers. After a short conversation, she puts the phone back on the table.

In this quite complex sharing of the phones, we assume that C is the owner of phone 1, B is the owner of phone 2, while A does not seem to have one at the moment. A therefore borrows phone 2 when she needs to make a phone call. C also borrows this phone to make a phone call, although she apparently had a phone of her own. The reasons for borrowing a phone while having one, might be, along with the argument in connection to excerpt 4, because she did not have "enough money on her phone", or because the phone call was somehow related to B, and she should therefore be the one paying for it.

At the end of the excerpt, the phone belonging to girl C, who is currently calling from her friend's phone, rings. Since C is busy talking when her phone rings, A answers. It is interesting that her answering is very quick, and she does not seem to get permission to get the call. There might be a previous agreement that A is allowed to answer C's phone, or maybe C subtly let A know that she could take it. Also, the caller identification function might be of relevance here. Perhaps A saw who was calling (or from whose phone someone was calling!) and thought that it was appropriate for her to answer.

Again, these examples of how the mobile phones are borrowed and lent within groups of friends show how the phone is not just a personal device, carried around and used by one person for that person's private purposes. Rather, it seems that the phone has become a collaborative resource among teenagers. If someone does not have access to a phone, that person can use someone else's phone. It seems that there is much trust involved here. They trust each other to borrow their phones, even when they are not in the same place and are therefore unable to see what the phones are used for. Showing someone the trust it means to let her or him use one's phone might be a way to display friendship; you don't let just anyone use your phone.

7.4.4 Sharing with Unknown Others

In all the above examples, the sharing of mobile phones, both concerning content and the actual phone, has involved teenagers who are friends or at least acquaintances that currently hang out together. We have seen how the phones are shared, wandering from friend to friend in intrinsic ways. As a contrast to this is the observation below, where the phones are handled by teenagers who seem to be unacquainted up until the described interaction.

> *Excerpt 8: Liseberg amusement park, Friday night*
> Two girls, about 15 years of age, are sitting on a bench. Next to the girls are some boys of about the same age. Two boys approach the two girls.
> Boy A: Are you Swedish?
> Girl A: Yes.
> They talk about their origin for a while. He insists on getting her phone number. During the entire conversation, he is holding his mobile phone in his hand. He gives the phone to the girl, and she enters something using the buttons (probably a phone number). Right after this, the other girl enters something on the phone belonging to the other boy.

BA: How old are you?
GA: Fifteen.
BA: Okay.
The conversation ends with the first boy asking the girl:
B: Is it okay if I call you at three o'clock tomorrow?
G: Yes.
And the two boys walk back to their other friends, without saying goodbye to the girls.

During this whole event the participants have their phones in hand, visible to each other. The mobile phones are here shared for social purposes, to make contact. The two boys use their phones as tools for getting acquainted (or flirting, it may seem) with other teenagers. For these boys, it is almost as if having a mobile phone in hand is an excuse for approaching the girls.

The example above also nicely demonstrates the ways in which the phones are used for social interaction in the local context. In this case no communication via the phone is involved. The phones are used for exchanging phone numbers with the group. Their meeting might lead to calling later (as one of the boys suggests), but at this moment all they do with their phones is going on right there, in the local milieu. Thus the phones are used for getting things accomplished in the local context, rather than calling non-present others.

7.5 Discussion

Our empirical studies of everyday use of mobile phones have examined the ways in which the mobile phones are brought into play when teenagers do things together. In this section we discuss the findings, as well as draw conclusions about what they might imply for design of mobile phones. First, however, we discuss some of the methodological issues raised in this study.

7.5.1 A Note on the Method

The material used in this study was collected through ethnographic fieldwork, and documented in field notes. This was sometimes a shortcoming. When analysing the data, many times we lacked some crucial piece of information, which we could not remember, had not written down, or simply had missed. For instance, sometimes we wanted to know how and where the phone was placed on the table after using it, but had no notes of this. In order to do a more detailed and comprehensive study of how the sharing was carried out, we would have needed video and audio recordings. For instance, in order to develop the notion of the turn taking of phones within a group, it would be necessary to analyse the conversation within the group.

However, it is important to remember that it would have been difficult if not impossible to get audio and video recordings of the natural occurring action described here. Studying the use of such a highly mobile technology as the mobile phone poses difficulty to the use of audio and video based analysis. It would perhaps have been possible in the cases where people were a little less mobile, such as the cafés, where we could have recorded a certain table, for instance. In other locations, e.g. the amusement park, we would have needed several cameras and microphones in quite a lot of places in order to collect as much material as we have done through observations. One possible solution would be to use mobile recording equipment instead of fixed, where the people observed can be followed around and recorded. This clearly raises many ethical concerns, at least when studying people in public settings. Having mobile equipment would imply that no specific place is used for collecting data, which makes it more difficult to inform people of where they might be recorded. Until these issues have been satisfyingly resolved, there is much interesting and useful data to collect through ethnographic observations. In relation to this, we want to stress the benefits of having two people in the field at the same time. This made it possible to collect and note more details of the interaction. We believe that we got considerably better material from doing fieldwork together.

Some of the issues could have been investigated further by using some sort of interview technique. Certain things are not possible to get at in a field study where no questions are asked. For instance, it could be useful to ask why they use their friends' phones when they have their own. However, these questions would have been particularly interesting to ask from the people we had observed, in connection to the particular instances in which the actual sharing took place. In the future, we are likely to see "traditional" methods for data collection develop to meet the issues raised when studying the everyday use of mobile technologies.

7.5.2 Designing for Sharing and Local Use

This chapter has described the local use and sharing of mobile phones. We have seen how the mobile phone has taken an important part in the local social interaction among teenagers. Teenagers do other things with their phones than just one person calling another, dislocated person. When teenagers spend time together, the remote communication they take part in, i.e. the phone calls they receive or make as well as the SMS messages, is rendered public in various ways, as described in this chapter. This is what we call strategies for sharing of content. The mobile phone itself is also shared between friends, borrowing and lending each other's phones. There was even an instance of sharing between unknown others, with the purpose of making contact.

The strategies for sharing presented here can be summarised briefly as two types: minimal forms of sharing and "hands-on" sharing. The *minimal form of sharing* of content, mainly SMS messages, was sharing by reading the message aloud, and sharing by showing the display to others. Both strategies are ways to let friends (and perhaps others in the surroundings) take part of or part in personal communication – ways to render private information displayed on a very small screen accessible to others. The *"hands-on" sharing* meant that the phone itself was shared between two or several teenagers. The hands-on sharing is made possible by the ecological flexibility of the phones; they are small, and easy to move around, thus enabling several people to use one as a collaborative resource.[8] This allowed for several teenagers to take turns in a phone conversation, pass around phones to share SMS messages, etc.

The local use of the phone, the "micro-mobility" in the words of Luff and Heath (1998), has many implications to a device originally introduced for remote communication. As pointed out by Luff and Heath, micro-mobility, defined as "the way in which an artefact can be mobilised and manipulated for various purposes around a relatively circumscribed, or "at hand", domain" (1998: 306), is a type of mobility not normally associated with new technologies. For those interested in designing new mobile phones and services, it might be useful to consider that phones are used within the local domain as well. The ways in which teenagers share the phones and their content today, could be drawn upon to add features or services that support this sharing. We will discuss a few possibilities here, based on our findings.

The sharing of mobile phones observed in this study raises questions about the notion of the mobile phone as a *personal* device, belonging to and being used by one individual.[9] This notion of a personal phone often goes hand in hand with it being perceived as a *private* phone (see for example, Chapter 9). With this follows the issue of the handling of private talk in public settings, something that occupies researchers (e.g. Persson, 2000) as well as the public. Of course, the mobile phone is personal and private in many ways, and is often used like that. We do not want to claim that these issues do not exist, or that they are not interesting to study. However, we hope to have shown with this study that the mobile phone can be something other than personal and private. Among the teenagers we have studied, the phones are not just treated as personal, and the calls and SMS messages are not treated as private. Rather, there is much work going on to render them *not* personal and *not* private. Understanding the ways in which this work is done, and what it involves, could be useful to consider when designing mobile phones. To a large extent the mobile phones of today are designed to be used by one person at a time. People using the phones for collaborative purposes have to work around this. One approach could be to make the phones more flexible, supporting the private, individual use, as well as the collaborative, group-use described in this chapter.

The extensive sharing of mobile phones has in a sense replaced the use of the public telephone provided by society. If you are without a phone, or without enough money on your phone card, or have other reasons for being "off-line", you can use someone else's phone. We have seen examples of how mobile phone owners very willingly, it seems, lend their phones to those without phones. In the cases described in this chapter, the mobile phone is a personal telephone made public, with the important difference that it is made available for selected people. The phones are shared between friends, or people who have (or want to have, perhaps) some relation to each other. This could probably be studied effectively by walking up to a stranger on the street and asking to use the person's mobile phone. Unless there were very good reasons, e.g. an emergency, this would probably strike people as odd behaviour.

One of the possible explanations of why phones are shared only with people with which one is acquainted could be economic. When using someone else's phone for calling out the owner is actually doing the caller a favour costing the owner money. A nice solution was reported to us by a young girl, who said that some teenagers who do not have phones buy prepaid phone cards which they can use on their friends' phones. The teenagers we studied did not seem to have great concerns about economics when lending and borrowing phones. Still, there could be ways of making the methods of payment more sophisticated, enabling someone other than the owner of the phone to pay for calls.

Another issue that the results from this study raise is the way teenagers spend time in groups, and how this is shown in their use of the mobile phone. In particular, our observations show how teenagers sometimes call someone's mobile phone and end up talking to someone else in the group. Sometimes they seem call a person up when looking for someone else, e.g. when A wants to talk to B and knows that B does not have a mobile phone, A calls to C's phone, assuming that they spend time together. The call is thus directed to a phone in that group, with the aim of getting hold of someone who might be a part of the group. This could be compared to calling a person's house, i.e. with a landline, looking for someone who does not live in that house but who might be there. There could be interesting technological solutions to this. Instead of guessing or having to know in advance who currently spends time with whom, there could be other ways of getting this information. Various awareness services could be built into the phones. This would of course raise some privacy concerns, which would have to be solved.

We believe that the mobile phone industry could learn from an understanding of the actual everyday use of the phones and services they are designing. It does seem that the industry is beginning to catch on to the young generation's use. Recently, there have been a number of new phones and services especially oriented towards young people. The mobile phone companies Nokia and Ericsson have both released mobile phones marketed in Sweden as "teenage phones". The industry focuses on developing further those services that seem popular. Different possibilities of using SMS are explored, for instance voice recorded SMS, multiparty SMS, etc. It remains to be seen how this will be accepted among teenagers.

Acknowledgements

The fieldwork conducted for this chapter was part of the iTeens-project, funded by the Mobile Informatics Program, SITI. We want to thank everyone who has commented on this chapter, among them Barry Brown, Jon Hindmarsh, Lina Larsson, Magnus Bergquist and Simon Love.

Notes

1. In particular, the Scandinavian countries, where we are based, have a high number of mobile phone users. Statistics for mobile phone usage quickly get outdated. However, some numbers might be useful to set the scene. In 1999, the European Commission's Telecommunications Watch reported a 58% mobile phone penetration in Finland, 51% in Sweden, 31% in Japan and 25% in the USA. Since then, the use in the Scandinavian countries and in Japan is said to have increased (for instance,

Wapland, 2000). The statistics for young people in Sweden tell of numbers around 75% among 15–17-year-olds. In Helsinki, Finland, the numbers from 2000 are as high as around 80% for 15 year olds, and around 90% for 16–18 year olds, the number being lower for boys (Keskinen, 2000.)

2. Christina Eriksson and Frida Norin (2000) "Female Application for the M-generation", unpublished master's thesis from Royal Institute of Technology (KTH), Stockholm.

3. Another note on the ethics of this study: in reporting our observations, all the material has been anonymised. We do not use any names that we have heard. We only describe the teenagers with an approximate age, from our estimations. Further, the place and time in which we made the observations are described in a general manner only, which should make it difficult to identify any single person.

4. With the exception that we were making notes.

5. Originally, we were interested in collecting both audio and video recorded data, but due to technical and ethical difficulties no such recordings were made.

6. We recognise that methods relying on recollection can be questionable; Sacks describes it as "very bad" (1992/1995, vol. II, p. 5).

7. She was riding on the tram with two girls, B and C. B is the only one described in the excerpt; she was sitting next to A on the tram, and was consequently closer to the phone.

8. To this extent, the mobile phone shares some (but definitely not all) characteristics of the various paper documents studied by Luff *et al.* (1992).

9. This notion is sometimes explicit in the literature about mobile phones. For instance, one writer calls it the mobile PP – personal phone (Roos, 1993).

References

Coogan K and Kangas S (2000) *Young people are the acrobats of communication.* Nufix-project, Elisa Communications, Helsinki. [In Finnish on http://www.alli.fi/nuorisotutkimus/julkaisut/nufix/]

Heath C and Luff P (2000) *Technology in action.* Cambridge: Cambridge University Press.

Keskinen V (2000) Ungdom och mobiltelefoni i Helsingfors [Youth and mobile telephony in Helsinki, original in Finnish and Swedish] Kvartti. *City of Helsinki Urban Facts Quarterly*, 4, pp. 17–25.

Kopomaa T (2000) *The city in your pocket: birth of the mobile information society.* Tampere: Gaudeamus.

Koskinen T (1999) *Mobile asynchronous communication: use and talk of use among a group of young adults in Finland.* Extended Abstract for the second workshop on HCI with mobile devices, Edinburgh, Scotland, May 1999.

Ling R (1998) "One can talk about common manners!": The use of mobile telephones in inappropriate situations. *Telenor: Telektronikk: no. 2*, 1998.

Ling R (1999a) "We release them little by little": maturation and gender identity as seen in the use of mobile telephony. *Telenor R&D Report 5/99*, Kjeller, Norway.

Ling R (1999b) "I am happiest by having the best": The adoption and rejection of mobile telephony. *Telenor R&D Report 15/99*, Kjeller, Norway.

Ling R and Yttri B (in press) "Nobody sits at home and waits for the telephone to ring": Micro and hyper-coordination through the use of the mobile telephone. In J Katz and M Aakhus (eds), *Perpetual contact: mobile communication, private talk, public performance*, Cambridge: Cambridge University Press.

Luff P, Heath C and Greatbatch D (1992) Tasks-in-interaction: paper and screen based documentation in collaborative activity. *Proceedings of CSCW '92*, Toronto, Canada. Seattle: ACM Press.

Luff P and Heath C (1998) Mobility in collaboration. *Proceedings of CSCW '98*, Seattle, WA: ACM Press.

Mitsuoka M, Watanabe S, Kakuta J and Okuyama S (2001) Instant messaging with mobile phones to support awareness. *Proceedings of Saint 2001: Symposium on applications and the Internet.* Seattle: ACM Press.

Persson A (2000) Intimitet bland främlingar: Om mobiltelefonsamtal på offentliga platser. [Intimacy among strangers: on mobile phone conversations in public places.] *Sociologisk Forskning*, 2.

Roos JP (1993/1994) *Sociology of cellular phones: the Nordic model*, Telecommunications policy. 17(6). Available on the internet at: http://www.valt.helsinki.fi/staff/jproos/mobiletel.htm.

Sacks H, Schegloff EA and Jefferson G (1974) A simplest systematics for the organisation of turn-taking for conversation. *Language*, 50(4), pp. 696–735.

Sacks H (1992/1995) *Lectures on conversation*, G Jefferson (ed.). Oxford: Blackwell.

Wiberg M and Ljungberg F (2000) Exploring the vision of anytime, anywhere in the context of mobile work. In *Knowledge management and virtual organisations: Theories, practices, technologies and methods*. Brint Press

Chapter 8
Running and Grimacing: The Struggle for Balance in Mobile Work

John Sherry and Tony Salvador, Intel Corporation

Good jazz is when the leader jumps on the piano, waves his arms, and yells. Fine jazz is when a tenorman lifts his foot in the air. Great jazz is when he heaves a piercing note for 32 bars and collapses on his hands and knees. A pure genius of jazz is manifested when he and the rest of the orchestra run around the room while the rhythm section grimaces and dances around their instruments. Charles Mingus

8.1 Introduction

The current proliferation of mobile devices in the computing, electronics and communications industries has begun to provide consumers and workers with types of experiences that once required being tethered to a particular piece of technology at a fixed location. Advertisers bombard us with images of executives reclining on sun drenched beaches, cheerfully pecking away at their laptops, or struting, like alpha males through airports checking their stock portfolios on PDAs and mobile phones.

In reality, new forms of connectivity require more careful attention to the types of behaviours incumbent on users than these utopian visions provide. Specifically, as we argue below, users of new mobile technologies must master, in an almost jazz-like way, a balance between what is going on "here and now" and the ongoing flow of activities that are outside the immediate physical context but which still require attention. Like the pure genius of jazz, these mobile professionals[1] running around the figurative room must remain ever more cognisant of their own respective rhythm sections, grimacing and dancing around their instruments at some remove. Unlike the pure genius of jazz, however, ordinary people do not always have the reserves of skill and experience on which to draw to achieve this balance. Without careful attention by designers – perhaps even in spite of it – those who are doing the running too often wind up being those who are doing the grimacing as well.

8.2 Intel's Architecture Labs and Mobile Technologies

Industry has witnessed a marked increase in the use of ethnographic methods over the past two decades, particularly in high tech (cf. Holtzblatt and Jones, 1993; Hughes *et al.*, 1997, Blythin *et al.*, 1997). At the People and Practices Research group (PaPR) in Intel's architecture labs, we have engaged in a variety of research projects which have impacted our perspective on mobility.

For the sake of brevity we will focus on two of the more recent projects, both of which have focused on mobile professionals as they conduct their daily work. The first of these studies focused on business travellers. In this project, conducted in early 2000, we shadowed and conducted *in situ* interviews with roughly a dozen individuals who were away from home five or more days per month. Sessions lasted anywhere from a half day to multiple days, and either involved actual travel with the individual between cities, or shadowing up to or starting from the airport gate. In this project, our concern was not to sample a wide range of variability, as our initial goal was to understand some of the basic issues involved with business travel. We did, however, make the effort to observe both men and women travellers, and to control for a moderate range of job types and incomes.

A second project examined in some detail the working lives of individuals whose mobility was more focused on particular urban or rural areas. This project entailed detailed observations – ranging from a half day to multiple days – of nearly two dozen individuals, at a variety of locations both in North America and in Europe. Shadowing protocols and *in situ* interviews were again employed in the course of this study. We also made an explicit effort to explore a range of variability. We stratified our sampling in terms of geography, for instance, by controlling for size of city (large urban centres, moderately sized cities, small and rural towns), Eastern USA versus Western USA (plus Vancouver BC), and heavy use of public transportation versus private vehicles. Sites included New York, San Francisco, Vancouver BC, Bellingham, Washington, Pella, Iowa, Albuquerque, NM, Stockholm, Sweden and Portland, OR.

In this project we attempted to sample a wider range of professions than in the business travel study. Those whom we followed included such diverse professionals as an event co-ordinator, a telephone line installer, a rural large animal veterinarian, a real estate agent, a physician/entrepreneur, an interior designer, a documentary film maker, and a variety of others whose daily work took them out of any office environments for prolonged periods of time.

In the midst of conducting these two ethnographic projects, we also as a matter of course gathered data on a third type of mobility – mobility restricted to a particular physical plant or office building. Few of the individuals we observed in the course of these two projects was "on the street" (on foot or in a vehicle) or in the air all the time. Most navigated a rather circumscribed physical space – a "home base" of sorts – for at least part of the time we shadowed them. Our analysis of this "in the building" mobility provided us with issues that at times complemented and at times closely paralleled the two more far-flung forms of mobility.

8.2.1 Understanding Mobile Work – Variable Environments

As mentioned, the emphasis here is on mobile work – thus a key to understanding such work lies in the acknowledgement that workers are traversing or occupying a variety of environments, rather than the (often taken-for-granted) environment of the desktop PC. We distinguished among a variety of elements of "the environment", including:

- The micro-environment: This refers to the immediate physical surroundings supporting or inhibiting technology use. This element pertains to such issues as what kind of furniture is available, whether the person is standing, sitting, walking, or in some other posture, what is the quality of ambient light, and other immediate physical effects.
- The physical "macro-environment": This aspect covers such issues as available infrastructure for power, connectivity, and other resources, mode of transportation, climate and geography.
- The temporal environment: This aspect deals both with available time and demands on both time and, interestingly, attention.
- The socio-cultural environment: This important aspect deals, most simply, with what is considered appropriate in a particular environment, relating – for instance, talking out loud, or using particular kinds of devices. It also deals with broader issues of cultural geography, for instance, the interplay between the temporal organisation of social life, urban geographies, and transportation infrastructures.[2]

In considering environments, we were careful to keep in mind that the environment is not simply a "container" in which activity happens. All the above elements interact in complex, often unpredictable ways. More importantly, human beings actively create their environments as much as they are shaped by them, either through the use of language to create the social environment (cf. Duranti, 1989; Hymes, 1972), or, as was common in the present study, by actively manipulating their own surroundings to accommodate their needs.

Our goal was to observe the interaction of the environment with the types of activities people engaged in, as well as the types of objects and technologies they used to accomplish these. Some of our insights came rather readily, for instance that laptop usage typically occurred in micro-environments that were remarkably similar to the standard desktop PC environment (as will be discussed below). Among other things, this led us to a distinction between two key elements of what is typically labelled "mobile" work: *remoteness*, which means separation from a resource-rich "home base", and *truly mobile work*, which involves both remoteness and motion, or at least more fleeting periods of stasis.

We were also able to identify some of the key activities that occupy mobile workers. As mentioned, we noted that all such activities were carried out with a sort of jazz-like attempt to continuously harmonise not only multiple activities, some of which are directly observable and some of which are not, but also to harmonise planned activities with improvised activities. One of the basic insights of our research (paralleled in other research – cf.Darrah *et al.*, 2000), is that there is a thorough mixing of planning and improvisation in the course of much of daily life, and particularly in the midst of mobility. We noted, for each of the following experience types, a thorough interplay between planned and improvised activity.

8.2.1.1 Logistics

The logistics of motion, which included moving self and stuff, locating and accessing needed resources and/or people to achieve this movement, and co-ordinating logistics of motion with others who are likewise in motion. Cancelled flights, overbooked hotels, and overtime meetings consistently force mobile workers to use all their skills to maintain a sense of harmony despite the need to improvise – to recover from the unexpected and reintroduce structure to their trip.

Consider the following example, extracted from an actual case of a consultant catching a flight from Portland to Chicago. In the routine mode, finding a flight is a (relatively) leisurely activity, where the customer typically has the ability to choose the most direct, the cheapest, or otherwise most desirable flight. In the exception mode, however, when a flight has been cancelled and there are few apparent alternatives, the traveller must improvise, take what she can get, and do so quickly before someone else takes the available options. Our consultant placed no fewer than 26 phone calls, leaving about a dozen voice mail messages and sending as many emails, in the attempt to get a new flight, notify colleagues in Portland of the change in plans, notify the potential customer in Chicago of the delay, and determine, after all, if it was even worth the trip.

In the routine mode, far fewer calls would have sufficed, because leaving a message and waiting for a response is often acceptable. In the exception mode, however, escalation was necessary, or, if that fails, one must try other numbers, or find an alternative contact, all of which the consultant did.

8.2.1.2 Dealing with Work

Beyond simple logistics, mobile professionals must attend to their actual *work*. Being remote from one's "home base" (which was usually the case for business travellers) implied a lack of access to resources, an inability to co-ordinate easily with others, and an inability to engage in ad hoc forms of communication and collaboration that routinely happen "at home". Not surprisingly, one of the key issues that most business travellers faced in this regard was connectivity. Many of these people lacked the ability to remotely access documents or information that resided on machines or databases "back at the office". This includes such simple resources as paper or Post-it Notes, but also, and most importantly, other people and access to technological support. Frequent travellers expressed a concern voiced also by telecommuters (cf. Schwartz *et al.*, 1999) – a sense of isolation in their work, and the lack of support for difficult-to-manage technology.

8.2.1.3 Managing Personal Affairs

Managing personal affairs constituted the third major type of experience we noted for business travellers. This included staying in touch with loved ones, taking advantage of the locality or situation occasioned by business travel, obtaining news, and managing household logistics. We were surprised in our interviews and observations of business travellers how frequently issues in this category emerged. They ranged from such simple needs as finding a decent restaurant near the hotel, to far more complicated issues, such as how to take care of a complicated family matter involving both finances and relationships, while far from home.

There were other, far more spectacular uses of technology to support personal life while on the road. One of our interview subjects – an avid diver – explained to us how, before every business trip, he visits a website featuring large amounts of detailed information about local shipwrecks, including the exact diving co-ordinates. This information was optimised for downloading onto a Palm device (in his case, a Handspring Visor), making it tremendously valuable to him while travelling, even when he didn't bring either a laptop or an internet connection (but always a GPS device).

8.2.2 The Jazz of Going Mobile

It was this combination of two realities – the need to harmonise among multiple flows of activity and the interplay of planned and improvised action – that led us to theorise the "jazz-like" nature of business travel. This was clearly supported, for example, in people's use of cell phones, travellers' workstations, palm pilots, assistants, friends, call centre personnel, public displays, and a host of other resources to seamlessly shift among their own work (calling a client to confirm or cancel a meeting), their personal lives (taking a call from a spouse, finding a gift in an airport), and the simple logistics of getting around (finding and catching a cab in a strange city).

While these insights originated with business travel, they extended to other forms of mobility as well. As we observed workers in a variety of "mobile" modes, we noted that considerable activity is inevitably devoted to maintain harmony between what is in the here-and-now, and what is not physically present but still demanding of our attention. This very mixing, as it turns out, requires skillful harmonising and improvisation precisely because it is fraught with tensions and potential conflicts. Before turning to the specifics, consider the following simple examples:

- In New York we observed a designer as she planned a party for the opening of a new office there. Her day consisted of logistics: planning the party and purchasing supplies, including alcohol, food, decorations, even plates, cups and napkins. She had no vehicle, but relied on taxis. She was in constant negotiation with possible vendors for shipping (she wound up carrying a case of champagne several city blocks). In order to expedite her chores, she delegated some of the shopping to a co-worker, which in turn introduced the struggle of how to monitor what she delegated (for instance, would the co-worker know to purchase the clear plastic cups, not the red ones). She also had to manage numerous other tasks, including interviewing a job candidate.

- In Stockholm, we followed an interior design consultant whose mid-morning doctor's appointment ended early, giving her just enough time to call her next client, tell her the doctor's appointment would end *late* (!), and squeeze in an extra meeting with a different client in the newly freed-up slot of time. This kind of opportunistic use of time formed a major type of improvisation on the part of mobile workers.

- Within the authors' own corporation, workers routinely hustle from building to building, attending a variety of meetings for project planning and reporting, for strategising, for gathering information, hearing reports on other teams' progress, and for making a variety of decisions regarding design, marketing or other issues. Meetings are a constant part of the workload of most group managers, and must almost inevitably be attended away from one's desk. Not long ago, ethernet networking jacks were made available in conference rooms, enabling owners of laptop computers with network interface cards to access the corporate internet from within the conference room. The original notion was that meeting attendees would be able to access meeting-critical documents or other electronic files they may have forgotten or neglected to carry from their desks.

Indeed, this is often the case. However, this new capability has also enabled meeting attendees to access their corporate email in-boxes as well – many of these employees receive dozens, even hundreds of email messages per day. Workers thus routinely use their network connections during meetings to stay abreast of email – a perfect example of the need to attend to what is not immediately present in the midst of an ongoing flow of activities in the immediate context. This practice has grown so widespread that jokes – and complaints – about it are common. Meeting attendees, distracted by email, fail to catch all of what transpires during meetings. No systematic study has been conducted that sheds light on how this affects the processing of such email. Clearly, however, what started with the intention of facilitating mobile work (even if that mobility is only "in building") has led to an escalation of the demands on workers by introducing two work activities, even two work environments, into a single physical space.

Initial evidence suggests that the problem is even more acute in firms that feature wireless LAN connectivity.[3]

8.3 The Technological Imperative: "Anytime, Anywhere Computing"

We introduce the following set of tensions by first introducing one that seems to lie near the heart of the matter – at least from the perspective of design and marketing slogans currently in vogue in the computing industry. Many of us are well familiar with it: mobile technologies are designed (or should be, we are told) to promote "anytime, anywhere computing". As we examined that objective in light of both what "computing" has come to mean in the desktop environment, and our ethnographic data, a fundamental tension appeared.

Consider the description of the desktop environment of a typical "knowledge specialist" as offered by Salvador and Bly (1997): individuals make use of heterogeneous channels of information, including a variety of forms of personal contact (both synchronous and asynchronous), a variety of media including the web, paper and locally stored electronic documents, and various idiosyncratic means for maintaining contact information, posting reminders, and a host of other things. The emergent picture is a sort of "command centre",[4] whereby a variety of media, channels, atoms and bits support what might be considered their computing environment.

The often voiced ideal of "anytime, anywhere computing" seems to ignore the value of this rich, command centre-like environment that supports much of the cognitive, social and communicative work associated with computing. Highly mobile workers hoping to maintain a high level of connectedness are faced with any number of complications associated with the fact that they occupy considerably different environments than their desktop command centres. First, many of the technologies routinely employed by Salvador and Bly's "knowledge specialists" were decidedly "old economy" – including paper, books and other "atoms" that extract high costs, primarily in terms of weight and convenience while travelling. In addition, even if much of the information required by highly mobile "knowledge specialists" managed to arrive in digital form, such workers usually do not have access to the kind of "micro-infrastructure" – furniture, multiple highly specialised devices, large displays, the ability to print, and places to physically separate various channels or objects – that allow them to easily organise their information. In some cases – for instance, sitting in a parked car with an open briefcase – it is possible to arrange a space in such a way as to organise (however temporarily) the flow of information into a coherent array. In other cases it is nearly impossible, although people continue to try.

The key contradiction in this situation is fairly simple: The technological imperative seems to be to provide as much information as possible to workers who are remote, in effect to provide them with all the information that they might enjoy in their "home environment". However, the environments in which they may find themselves often prevent them from organising this deluge of incoming information in the way they could at their home environment. Mobile professionals simply cannot easily deal with the information they routinely process at their desktops.

8.4 Access to Other People

This "anytime, anywhere" tension was particularly evident with respect to the ways mobile phone technology has affected contact among people. In many places, for instance in many large European cities, mobile phone technology is so widely deployed among working age adults that for all practical purposes it can be considered ubiquitous. This ubiquity presents an interesting design challenge – not just for technology creators but for those actually using it in their daily work. Because everyone knows that everyone else has a mobile phone, it is no longer part of the calculus of human contact that one is inaccessible by virtue of being "away from the phone". From the (somewhat simplistic) point of view of information sharing, this seems desirable from the perspectives of productivity and efficiency. Townsend (Chapter 5) among others, has called this an increase in "urban metabolism". Yet when "unfettered access" means the possibility of being interrupted by anyone at any time, the potential disruptions to work and concentration may outweigh the benefits.

Thus, a set of competing agendas arise. Knowing that others have access to phones, people expect that colleagues, friends or family should be accessible at any time, and in many cases desire easy and immediate access. At the same time (and as a direct result) people need a way of controlling access to themselves under the variable circumstances of mobile work.

This issue was clear, for instance, among virtually all of the individuals we shadowed in Stockholm, Sweden (one city featuring near ubiquity of cell phones). We were fascinated to watch one physician, who was also involved in a variety of entrepreneurial concerns, filter access to himself through an elaborate system of profiles and device configurations. His cell phone supported a number of different hardware configurations (e.g. ear piece plugged in, phone on belt; phone in charging jack in car; phone in charging jack at home, etc.). He accessed the website of Telefia (Sweden's national telecoms firm, privatised in the summer of 2000) to set up and categorise his many social and professional contacts into a system of profiles. At the same website, he then correlated particular hardware configurations with specific caller profiles to permit filtering. Physicians with whom he closely collaborated, for instance, would always have access to him during certain hours of the day, while others might have access to him only when he was outside of his clinic setting, but still during working hours. It was a complicated management burden that, he admitted, worked only partially, as the category mappings were somewhat rigid, the mapping between device configuration and categories was not entirely motivated, and there was no way to handle temporary or unanticipated exceptions.

We have noted a variety of other ways by which individuals will go to extremes to limit access to themselves, yet want to be "in touch" as much as possible on their own terms. Some individuals leave their cell phones perpetually turned off except when making calls. Others wrestle with email in-box management schemes that are every bit as complex as the phone filtering scheme described above. In every case, the basic tension was evident between uninhibited access out while filtering access in.

8.5 Awareness and Co-ordination

A variation on the "access to others" theme played out in terms of awareness and co-ordination. Examining a wide variety of issues faced by mobile workers, we theorised that a large number of these fit into problems specifically associated with "remoteness" – that is, physical or temporal separation from co-workers, work objects and other resources necessary for the completion of work. The framework proposed by Churchill and Wakeford (Chapter 11) provides some interesting insights in this regard. A variety of technological solutions, particularly in the CSCW community, have been designed to solve the problem of "tight coupling" – collaborative authoring environments, meeting environments, shared workspaces, etc. Less research and technological experimentation has focused, however, on looser couplings, what we identified in our work as "passive awareness and co-ordination".

For instance, our research suggested that cell phones were heavily used (both in Europe and the USA) as a tool for "checking in" – brief conversations that apparently serve the primary function of making sure both parties are "OK", along with some brief discussion of status or progress. This aspect of co-ordination is particularly interesting. As research has shown (cf. York, 1999), telecommuters (and, we propose, mobile workers) can suffer from a sense of isolation – they feel cut off from the rich potential for interactions in their office environment. We were particularly interested in exploring how to extend this sense of "passive awareness" beyond what the cell phone currently provides.

As mentioned above, however, this "OK-ness checking"[5] function is limited by the fact that it requires the active participation – however brief – of all concerned parties. This presents the problem of interruptions and intrusions. A seemingly natural extension to this capability, one that would apparently solve the interruption problem, lies in making this shared sense of awareness even more passive. For instance, by combining phones or wirelessly connected PDAs with buddy lists, location-based services, and perhaps some intelligent monitors on various aspects of a user's status, one could potentially create a set of mobile technologies that would interlink a group of people in a web of mutual passive awareness (McGuire, 2000).

The desire for something like this was explicitly discussed both by many of the people we observed and a variety of designers (Dourish and Bellotti, 1992). It seems to be a rather natural and desirable extension of what's possible in the physical workplace environment, where workers often enjoy a sense of the activities, status or at least presence of others based on the fact of their co-presence. However, to technologically implement this sense of passive awareness immediately introduces tensions of its own.

This basic tension has emerged in a variety of similar contexts. As experiments with active badges (Harper, 1992) have shown, the value of knowing what others are up to directly competes with the costs of giving up one's sense of autonomy and freedom from surveillance. Relationships of power weigh most heavily in this regard, as tools for passive awareness resemble all-too-closely the instruments of surveillance, marked, as Foucault (1979) has explored, by the hiddenness of the powerful behind instruments of scrutiny over the subjugated. At one interview, a contractor (who, notably, was himself in a position of management over half a dozen work crews) explicitly and succinctly rejected any strong form of passive awareness: "I need a way to check in with my guys, not spy on them." Similar examples also appeared outside the immediate workplace setting. We were told during one interview of an attempt by a group of parents (most of whom worked for a high tech firm) to set up web cams on the premises of a day care centre attended by their children. Not surprisingly, the suggestion met with considerable resistance by the day-care staff.[6]

8.6 Creating Information

Creating information in digital form – one of the most under-supported activities for many mobile professionals – provides yet another example of a basic tension among competing values. We were struck, in examining the activities of all kinds of mobile professionals, by the ubiquity of rather rudimentary paper management schemes, and the absolute lack of decent equipment by which to accomplish, in digital form, such simple tasks as taking notes, sharing lists or scrawling brief messages. In fact, there are currently no devices with any kind of mass distribution that fill this seemingly significant void between the personal digital assistant and the laptop. The former, we were told by a large majority of our subjects, is simply too difficult to use for any kind of data entry beyond contact information, passwords or calendar entries. The latter requires long periods of time to boot up, weight to carry, the inappropriateness of using a large, rather noisy machine in the middle of a variety of social situations, and, of course, the financial costs.

A number of vendors currently supply tablet-like devices, or will shortly.[7] The key to success here seems to be providing an experience that allows more, easier and richer input than current PDAs afford, but without the costs of space, boot up and usage time, and weight that makes laptops undesirable. While a richer interaction paradigm, greater screen real-estate, and easier data entry all seem to be promising opportunities for a highly mobile device, the addition of such features may result in costs that people are not willing to bear.

This issue is complicated by the fact that competing agendas seem to be at play in the technology industry itself. On one hand, evidence seems to suggest a convergence of functionalities onto single devices. For mobile professionals, the most widely discussed example seems to be the convergence of the PDA with the phone, which some analysts argue will inevitably merge in the USA and in Europe (Ward, 2000).

At the same time, a variety of technology offerings suggest that an opposing trend is also in play: we note what might be called a trend towards appliance-like devices, whose purpose and usage models are clear and intuitive. Furthermore, new connectivity technologies – Jini, Bluetooth, and others – promise to lead to at least some disarticulation of the elements of computing. Data storage, software, security, displays, and any number of elements of the computing experience might be removed from the context of the immediate device and placed elsewhere in the system – down the "wire", embedded in the environment, or attached to the person or immediate surroundings in new ways. This implies a world where people will take a more "architectural" approach to the choice of objects, services and other offerings.

But this future possibility raises as many questions as it answers: At what point does this proliferation of highly specific devices become counterproductive? If people are to take an architectural approach to their technological selves, can industry be counted on to provide devices and services that actually co-exist peacefully and productively? Current evidence from the PC industry, which has been plagued by problems of interoperability despite a situation of dominance by a few providers, suggests that this may not be the case. At what point, then, does the management burden of architecting a workable solution prevent the vast majority of users from spending the time and effort required to devise a satisfactory array of technologies?

8.7 Conclusion

Underlying these dilemmas is perhaps a single fundamental concern. Beyond the question of whether or not a given constellation of devices will actually operate, or the ways in which people access others easily while shielding themselves from interruptions or scrutiny, lies the question of how to balance "what is here" with "what is not". Some have suggested that digital technology's greatest strength is that it "allows you to do things that are free of time and space".[8] The problem, however, is that doing things that are free of time and space can be highly disruptive when the present time and space require our attention. Technology bursts into the here and now – it introduces entirely new worlds to which the mobile professional must attend, over and above the ones they currently occupy. It can call forth not just one rhythm section, grimacing and dancing around their instruments, but many of them, each with potentially different and interfering rhythms of their own.

The evidence of this disruption, the demolition of old boundaries, and the attempts to construct new ones, is everywhere. Palen *et al.* (2000) have pointed out how cell phone conversations create dual spaces which must be managed, separated, and sometimes joined. A variety of researchers (Schwartz *et al.*, 1999) discuss the (often hidden) work of managing boundaries created by the intrusion of work into personal lives – a situation only exacerbated by technology. At a very pragmatic level, research strongly suggests that certain types of technology usage while driving pose a serious safety threat (NHTSA, 1997).

In thinking about designing for this reality, it might be useful to distinguish between two types of mobile functionality: what might be called technologies of "transport" and technologies of "connectivity". PDAs bearing contact information, laptops with spreadsheets that have been downloaded from a server, or even the Visor with the diving coordinates of shipwrecks, all fall into the former class of devices that carry information that users have, at some point, loaded for the express purpose of future access on the go. These technologies were largely exempt from many of the tensions described above. The real tensions, it seems, result from connectivity. Cell phones, wirelessly connected PDAs, and email, despite different levels of synchronicity and intrusiveness, are all technologies of connectivity, and all carry more threats in bringing together what is physically present with what is not.

This is not a totalising dichotomy, of course. In a few cases we observed, people still had problems dealing with all the information on their laptops in remote work environments even when not connected. Still, the general trend seemed to be fairly clear. As so much effort in the mobile technology industry seems to be for the provision of connectivity (wireless PDAs, WAP, Bluetooth, wireless networking), it seems worth noting that it is this very technological capability that introduces such tensions.

So, what can designers do about it? That answer is not easy – perhaps they can do nothing on their own as a prime driver of a solution in this regard. Think of the simple, rather subtle social conventions that have sprung up around cell phone use, including the common question–answer pair:

> Q: Where are you?
> A: I'm on the train.

This simple adjacency pair has been widely deployed in mobile phone conversations as a way of setting expectations about an interaction's tone or duration, establishing what types of conversation are possible, and other meta-communication, based on a simple exchange of the respective physical contexts of the participants.

It is doubtful that designers could have anticipated either the need for – or the emergence of – this widespread communicative trick. But now that it is so evident, design can support or extend similar conventions in other ways. Designers at both Intel and Fuji-Xerox Palo Alto Laboratories have independently arrived at the notion of functionality on board the cell phone or other handheld device, allowing users to manage incoming calls flexibly, without actually having to answer the phone or speak. With a simple innovation, the phone relieves some of the burden of clashing worlds in a flexible way at little cost to the owner.[9]

Are such innovations readily apparent in other domains, for the various other tensions described above? A few rather obvious suggestions come immediately to mind, including placing a strong emphasis on shielding users from unwanted intrusions of non-critical information ("spam"), and ensuring that systems of passive awareness are implemented purely on the basis of mutual consent among peers. For each of these dilemmas a more thorough examination of both costs and benefits to users must be undertaken with specific attention to the objects at play and certainly with as much knowledge as possible about the environments of use. For some of these tensions, it remains to be seen whether they constitute constraints that can ultimately be satisfied by a technological offering, or whether they are inherent contradictions of mobile work that no technology can solve.

Notes

1. We use the term "mobile professionals" in its loosest possible sense, referring to any individuals whose work activities require mobility away from a desktop or other fixed location.
2. These factors – specifically, the fact that there are many social hubs in London, the fact that people have long commutes on public transportation and thus may not go home immediately after work, and the fact that homes are small and may not feature room for a personal computer – have all been identified as influential on the success of the EasyEverything cyber-cafés in London (Bell, personal communication).
3. Based on personal communications from some of our interview subjects.
4. Term of the present authors, not Salvador and Bly.

5. Technologists often refer to this activity as "pinging".
6. Staff were primarily opposed to being under the one-way scrutiny of parents. Concerns about security – who could access the web page – also surfaced during discussions of the web-cam idea.
7. As of the end of 2000, a variety of firms had offerings or announcements in the "tablets" product category (denoting a device somewhere between a PDA and laptop in size, input modality, etc.). But as of this writing there was no clear market leader, or, for that matter, any large market.
8. Statement made, for instance, by Nicholas Negroponte, *21st Century: World Without Walls*, Part III: "Culture in the Communications Age". Philadelphia: University of Pennsylvania and Koppel Communications, Inc. 1990.
9. For more details at Intel, contact Wendy March, interaction designer in People and Practices Research. Wendy.March@intel.com. See also the work of Les Nelson on "Quiet Calls" for a very similar problem formulation and solution, based on a simple three-button array rather than a thumb-wheel. http://www.fxpal.xerox.com.

References

Blythin S, Rouncefield M and Hughes J (1997) "Never mind the ethno" stuff, what does all this mean and what do we do now ethnography in the commercial world. *Interactions*, 4(3), pp. 38–47.

Darrah C, English-Lueck J and Freeman J (2000) *Living in the eye of the storm: Controlling the maelstrom in Silicon Valley*. http://www.sjsu.edu/depts/anthropology/svcp/ SVCPmael.html

Dourish P and Bellotti V (1992) Awareness and coordination in shared workspaces. In J Turner and R Kraut (eds). *CSCW '92: Sharing perspectives*, pp. 107–114, NY: ACM Press.

Duranti A (1989). Ethnography of speaking: toward a linguistics of the praxis. In FJ Newmeyer (ed.). *Language: the socio-cultural context*, pp. 210–228. New York: Cambridge University Press.

Foucault M (1979) *Discipline and punish: the birth of the prison*. New York: Vintage Books.

Harper R (1992) Looking at ourselves: An examination of the social organisation of two research laboratories. In J Turner and R Kraut (eds). pp. 107–114. *CSCW '92: Sharing Perspectives*. New York: ACM Press.

Holtzblatt K and Jones S (1993) Contextual inquiry: a participatory technique for system design. In A Namioka and D Schuler (eds). *Participatory design: principles and practice*. Hillsdale, NJ: Lawrence Erlbaum.

Hughes J, O'Brien J, Rodden T, Rouncefield M and Blythin S (1997) "Designing with ethnography a presentation framework for design": *Proceedings of the ACM conference on designing interactive systems: processes, practices, methods, and techniques*, 1997, pp. 147–158.

Hymes D (1972) Models of the interaction of language and social life. In J Gumperz and D Hymes (eds.). *Directions in sociolinguistics: the ethnography of communication*, pp. 210–228. New York: Holt, Rinehart & Winston.

McGuire D (2000) FTC talks wireless privacy. *Washington Post*, December 12. http://www.washtech.com/news/regulation/5935-1.html

Nippert-Eng C (1996) *Home and work: negotiating boundaries through everyday life*. Chicago: University of Chicago Press.

National Highway Traffic Safety Administration (NHTSA) (1997) An investigation of the safety implications of wireless communications in vehicles (online: http://www.nhtsa.dot.gov./people/injury/research/wireless (August 2001).

Palen L, Salzman M and Youngs E (2000) Going wireless: behaviour and practice of new mobile phone users. In *CSCW 2000*, pp. 201–210. New York: ACM Press.

Salvador, T and Bly S (1997) Supporting the flow of information through constellations of interaction. In J Hughes, W Prinz, T Rodden and K Schmidt (eds). *ECSCW '97*, pp. 269–280. Dordrecht: Kluwer.

Schwartz H, Nardi B and Whittaker J (1999) The hidden work in virtual work. Paper presented at the *International Conference on Critical Management Studies*. Manchester, UK, 14–16 July.

Ward M (2000) The future is at hand. *BBC News on-line*, 27 November. http://news.bbc.co.uk/hi/english/sci/tech/newsid_1042000/1042914.stm

York T (1999) Telecommuting causes, solves problems. *IDG/Infoworld*, June 2, 1999. http://www.cnn.com/TECH/computing/9906/02/telecommute.idg/index.html

Chapter 9
Blurring the Boundaries: Cell Phones, Mobility, and the Line between Work and Personal Life

Diana Gant, Indiana University, USA and Sara Kiesler, Carnegie Mellon University, Pittsburgh, USA

9.1 Introduction

Up until the beginning of the twentieth century, most people lived in close proximity to the places they worked – on farms, above stores and cafés, in back rooms of schools, and in boarding houses. Co-workers were often members of the family, companions and neighbours. With the rise of modern technology – electrification, motorised transportation, communication systems – and the growing importance of the bureaucratic work organisation, the separation between work and personal life grew more definite. Commuting to work, strictures against "personal calls" at work, socialising during weekends, and having a separate "personal" or social life, are twentieth century concepts. These concepts reflect differentiation of the social meaning of places and locations – working with other employees at the office versus seeing family at home, for instance. They also reflect differentiation of the social meaning of time – the 9 to 5 workday versus the weekend. But today, wireless technologies, which help people cross space, time, activity and social networks, promise to bring us back to earlier times when the boundary between work and personal life was less distinct, and to influence the meaning of space and time. This change is somewhat of a paradox, however, as wireless technology will also take us further afield, as it increases our temporal and spatial mobility. In this chapter, we discuss how this paradox is unfolding and draw on data from a field trial of digital cellular telephony to show some of the social implications for how we work and live.

9.2 The Social Meaning and Influence of Settings

Research in the tradition of "social ecology" (Barker, 1968) has shown that different settings for social behaviour – offices, meeting rooms, homes, restaurants, parks – sharply affect the way people act and the expectations they have of others. Mr Smith's behaviour in a meeting room and in a bar are likely to differ far more across these two places than Mr Smith's behaviour in the meeting room as compared with Mr Brown's behaviour in the meeting room. People derive the social meanings of different places, or *behaviour settings*, from a myriad of cues – their architecture and artefacts, technology used, rules in force and how most of people dress and behave (Cialdini *et al.*, 1990; Hatch, 1987; Proshansky *et al.*, 1970; Sproull and Kiesler, 1991; Stokols, 1990). These social meanings also influence people's mental schemas or models of different places and even provide scripts for what people say; more generally, they help organise people's social and work experiences (Edney, 1976). Sharing behaviour settings and acting in accord with the norms of these settings contribute to group identity, and increase people's satisfaction with their groups and their work (e.g. Newman, 1972; Baum and Valins, 1977; Edney and Uhlig, 1977). People in contiguous and similar behaviour settings tend to interact and to like one another (Moreland, 1987).

Temporal cues contribute to differentiation of social behaviour and mental models in different settings. For example, Mr Smith is less likely to stand around chatting with neighbours on a weekday than on a weekend. Temporal cues with social meaning include not only calendar and clock times, but also temporally bounded social practices such as "lunchtime". Technology has contributed to changes in the meaning of time. The dissemination of the electric light extended the "daytime" and increased the likelihood that both work and personal communication would move into the night and into less traditional locations (Melbin, 1978). More recently, the advent of computer networking, email, and the web has led to a widespread increase in after-hours work communication at home, and probably, to shorter deadlines in distributed group work as people expect faster response times.

The more legible are the cues that delineate different places, locations and times, the more clearly differentiated are they as behaviour settings and the better norms can be conveyed and obeyed (Gibson and Werner, 1994; Lynch, 1960). Consider whether airport vans and restaurants at the lunch period are appropriate for work or personal interactions. The legibility of the cues in these places is less clear, and hence less influential, than the cues associated with meeting rooms at 9.30am. Vans and restaurants usually lack office furnishings, wired telephones, clocks and other accoutrements signifying "workplace", but on the other hand, people dressed for work may populate them. Hence, although we inhabit many distinct behaviour settings, such as offices, that influence behaviour strongly, at the same time we also inhabit "mixed use" places such as restaurants, cars and public places, where the work and non-work cues may be less legible and implications for behaviour less clear.

Just as yesterday's technologies did, today's new technologies are contributing to changes in the social meaning and use of different behaviour settings, and to changes in the meaning and legibility of cues for behaviour settings. Wireless technologies (cellular phones and other wireless devices) embody a number of "affordances" (Norman, 1990) that suggest powerful social impact. First, these technologies reduce the costs and effort in communication with others; it has long been known that reducing the costs and effort of communication dramatically increases the likelihood of communicating (Zipf, 1949; Allen, 1977; Kraut *et al.*, 1987). Second, because they are not tethered to location, wireless devices that permit communication and access to information are likely to increase personal mobility and to reduce the constraints of location on where people are at different days and times. Third, since they operate independent of location and the control of household or organisation, these technologies are likely to reduce the constraints of time. The implication of these changes is a blurring of the cues and social meanings that separate our settings for work and personal life.

9.3 One-person, One-number: The Trial of a Personal Phone

We explore these issues drawing on data from the 1993 technical and marketing trial of a new digital (PCS) cellular service carried out by the former Bell Atlantic Mobile organisation. The "one-person, one-number" service was somewhat more advanced technically than current cellular services, since it gave participants in the trial the ability to make and receive calls using one telephone number whether on their wired phones or wirelessly. When users were away from home or office, they would use their "personal phone" with its personal phone number. But when they were at home, and someone called on the personal number, the participant's wired home telephone would ring. Similarly, when the participant was in his or her office, the office telephone would ring when the personal number was dialled.

The technology used in this trial automatically forwarded to the least expensive service and clearest connection, and gave users the ability to send and receive calls anywhere, including non-local areas. The main technical difference between the cellular phones used in the trial and current cellular phones was that current phones are slightly smaller. The main technical difference between the services offered in this trial and current cellular services was the ability, in the trial, to use one phone number, and to use one's wired and wireless services interchangeably. In addition, participants could customise their network services in great detail, e.g. forward calls from the home number to the cellular phone at certain times of day and not others. However, participants in this trial generally did not use these options, and used their trial phones much as people do today. Indeed, the lack of interest in one-number service by our participants (and by participants in other trials, we suspect) may have led to its withdrawal from the marketplace. Hence, this trial is a fair representation of the cellular technology and services we see today. A somewhat more detailed technical report on the trial of this technology is given in Kiesler *et al.* (1994); see also Hinds (1999).

This study was performed before the spectacular growth and dissemination of cellular phone subscriptions to individuals. (In one decade, 1990–2000, subscriptions grew from a few to one hundred million.) The comparatively early (1993) year of this study in the life of the cellular industry gave us a chance to explore this wireless technology before it had become a normal facet of everyday life, and therefore to observe the first effects it would have on social life (see Marvin, 1990). At the time of the study, we evaluated the affordances of personal phones, and set out to see how their use would increase communication, mobility and the likelihood of work and personal communication in work and non-work settings.

Trial participants were enrolled through work departments at Carnegie Mellon University to allow us to observe how the entire work groups responded to the technology. Three departments – Information Technology, Public Relations, and Theatre Production – containing 25, 23 and 10 participants respectively, took part in the study. Information Technology develops and maintains the computer, networking, and telecommunications operations all over campus. Public Relations manages communication with the mass media, community affairs and external publications. This department had staff in 12 different buildings. Theatre Production provides support and supervision of students in costuming, set design, electronics, lighting, woodworking and other activities related to theatre and television productions. The department put on many productions that required staff to travel off and on campus to acquire materials and supplies.

Data collection occurred during a 10-week period beginning in late January and ending in early April of 1993. Data collection proceeded in three phases, beginning on a staggered schedule one week apart for each of the three departments. In the first stage, we conducted a survey and asked all participants in the department to keep a log of all their communications and locations (including face to face) for an entire day. Everyone was interviewed the day after this communication diary. Then, the personal phones were distributed to all participants. Base stations were installed in their homes and offices, and in some cases, their cars were wired for hands-free communication. In the second stage, we conducted another survey and an audit of how participants were using their personal phones. In the third stage, we asked all participants to keep a log of all their communications and locations (including face to face) for an entire day. Everyone was interviewed the day after this communication diary.

9.4 Results

At the outset, we must explain that participants had considerable difficulty, at first, using their personal phones. The problems they encountered included the following:

- Turning the phone on and waiting for it to work (not understanding that one must "send" a connection), and hanging up (need to "end" the call).
- Figuring out how to set up and obtain access to voice mail.
- Interpreting voice prompts.
- Navigating menus and modes, especially going backwards.
- Clearing the display when making a dialing error.
- Understanding the concept and use of network customisation options, such as call forwarding and call waiting.

To overcome these difficulties, participants frequently asked for help from one another, the researchers, and the technical staff associated with the trial. Pamela

Hinds (2000) ran an experiment on the voice mail feature that illustrates the difference between what our participant-novices actually experienced and what the cellular sales force and support staff thought new subscribers would experience. These experts underestimated it would take participants only 13 minutes to learn to use voice mail, following the manual and a one-page list of "simple" instructions. Yet it actually took between 20 and 23 minutes for trial participants to learn voice mail, a difference of over 50%. Indeed, many participants ended up never using voice mail, and practically none of them used any network customisation options.

In spite of these difficulties, the overwhelming majority of the participants quickly learned to use their "personal phones" in basic mode, were extremely enthusiastic about them, and used them to make and receive on average about 12 calls per day. Participants, after receiving their new personal telephones, became more mobile. That is, they spent more time in locations away from home and office and communicated in more mixed-use settings such as hallways, homes, cars, restaurants and outside. They received proportionally more communications in these places as well. One participant took a call from the president of the university while seated on a toilet!

Participants were thrilled with their ability to be mobile while communicating. One mother of a disabled child, who had had to be near a wired phone in case her child's school called, was now able to leave her office and home. She said, "It changed my life."

As they carried their personal phones from place to place, participants formed a sense of their personal phones that was quite different from their sense of their wired telephones. At one point, we asked those who might not need these phones very much to contribute them to someone else; we got only four volunteers, and messages such as the following:

> ...I love my personal line phone. I think the trial should continue for another year.
> ...I need it ... you won't get a volunteer in me.
> I LOVE MY PHONE!!! Bell Atlantic will have to tie me down to get it back ...
> My phone has become a permanent part of my anatomy....

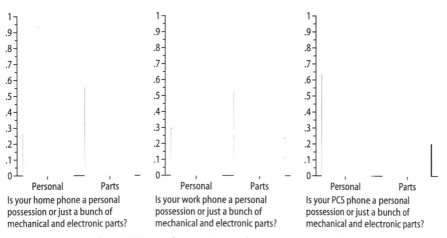

Figure 9.1 Personal possession or electronic parts questions

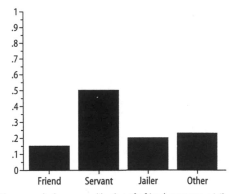

Is your relationship with your work phone more like that of a friend, a servant, a jailer, or something else?

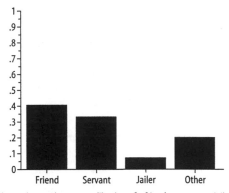

Is your relationship with your home phone more like that of a friend, a servant, a jailer, or something else?

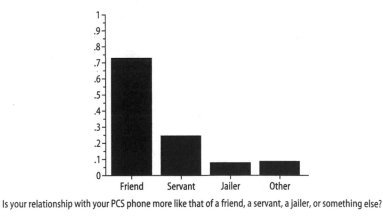

Is your relationship with your PCS phone more like that of a friend, a servant, a jailer, or something else?

Figure 9.2 Friend, servant or jailer

On our survey, we ask participants about their "relationships" with all their telephones. The stark results are shown in Figures 9.1 and 9.2. In brief, people thought of their little personal phone as a possession, an attitude that was very different from their attitudes about their wired telephones.

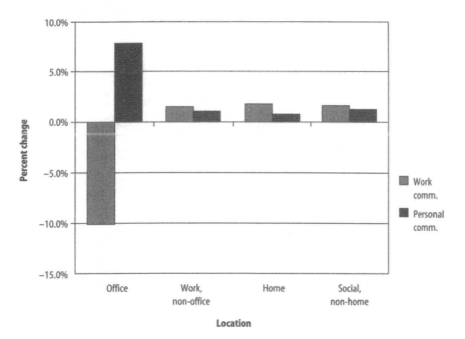

Figure 9.3 Change in location of communications received after cell phone introduction

With more mobility and a changed concept of what it meant to "have" a phone, the nature of participants' communications changed (see Figures 9.3 and 9.4). Participants initiated and received more personal communication in transitional work settings such as hallways and lobbies. Many said they were able to conduct personal business during otherwise "dead" communication time. For example, participants reported calling family while walking across campus – a work setting, but one where no prior phone access existed and where the personal call was not displacing ordinary work. We did not find that employees initiated substantially more personal communication from their internal offices, where the behavioural norms were very strong. Rather, the personal phones served to further weaken the salience of work norms in settings where the norms of communication were already weak.

On the other hand, participants did receive more personal communication in strong work settings. The "one-person, one-number" feature of the technology circumvented the strong social cue associated with the workplace phone number. For instance, the child of a trial participant called her mother at work on her personal phone and interrupted a meeting. The child did not normally call her mother at work for non-critical conversations, but since it was her mother's personal phone and not her work phone she though it was OK to do so.

This is not a one-sided story. Trial participants also initiated more work communication in non-work, personal and social settings. Unexpectedly, however, the main increase in work communication occurred not via the personal telephone, but face to face. That is, participants exploited their increased mobility to meet

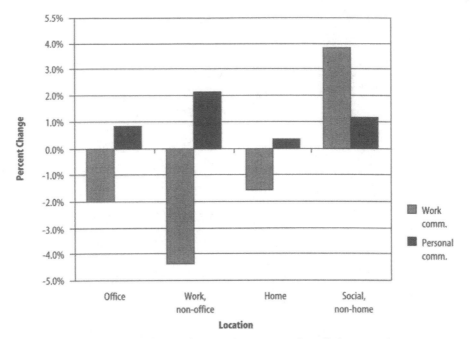

Figure 9.4 Change in location of communications sent after cell phone introduction

with co-workers in cafés, restaurants, parks, and homes to discuss work. The personal telephone gave them the ability to leave the office without worrying about missed phone calls. This was particularly true for employees in the Theatre Productions department. The personal telephones allowed them to shop for props together and still stay in touch with the office.

The "one-person, one-number" technology also made participants more susceptible to work-related phone calls at home. In one case, the supervisor of an employee in the Information Technology department called the employee in the evening with a technical, but non-emergency, question. The boss knew that the employee was off duty, but figured that he would call anyway. The boss had never called this employee at home using the employee's home telephone number.

To evaluate how these changes in communication might be related to temporal changes as well, we evaluated work communication during contractual work hours and non-work hours. To cover all the employees' work hours, we defined work hours as 8am–5pm excluding the noon lunch hour. Before the trial, 61% of all communications occurred during work hours. At the post-test, there was a small decline – 56% of all communications occurred during work hours. This result suggests more of participants' communications happened outside of work hours when they were using personal phones. We then re-estimated the regression logit models using only work hour communication to test the possibility that the increase in social communication in work settings might have happened due to people spending more time at the office during non-work hours. Results indicate that this was not the case. In fact, social communications sent in work settings during work hours increased from 9% to 11%. Also, social communication received in work settings during work hours increased from 9% to 11%. These are not statistically significant changes, but do suggest that people tended to be less bound by the temporal cues of "work hours".

9.5 Conclusions

Many new information technologies intentionally change our work and lives. Wireless technologies were long intended to increase the efficiency of distributed and mobile business. Our research addresses some of the unintended social effects, and perhaps unexpected effects, of this technology. When we began our field trial of personal PCS phones, cellular telephony had not yet taken the USA by storm; our first surprise was how much our trial participants used these phones for both work and social purposes despite the considerable usability problems they faced. We learned that trial participants had become personally attached to their new phones and thought of them as personal possessions. Our results showed a clear shift in work and personal communication in behaviour settings. Our data showed that participants were sending and receiving personal calls in work settings and work calls in more social, personal settings. Telephone calls on the personal PCS phones, on the whole, were more spontaneous and unrelated to place and time than were calls made from wired telephones.

Many participants had mixed feelings about these changes in their own and others' behaviour. In interviews they said they wished they had buffers to prevent unwanted calls. Indeed, they avoided using the one-number option because they were wary of giving every potential caller the same telephone number. They were not as eager to be buffered by others, though. And only 2% of the participants used the call screening services offered. It is possible that participants' common behaviour of answering all their calls was associated with the novelty of the technology or with usability problems. Perhaps with more experience and design changes, participants' behaviour would have changed. Perhaps they would have blocked more social calls in work settings, for example, and left more calls for voice mail. Our data speaks only for the initial months of usage, not for long-term changes in behaviour. However, arguing against this possibility, we found that the IT professionals in the study showed an even greater blurring of boundaries of place than other participants did, and these were the very participants who experienced few usability problems and for whom cellular technology was less novel. And too, our observations of cellular use everywhere we go suggests that these phones are often used spontaneously without regard to time or place. The blurring of many old boundaries separating work, social life and personal life seems already widespread.

We believe a paradox of wireless technologies (cellular phones and other wireless devices) stems from their low cost and ease of use (in "plain-vanilla mode"), from their portability, and from their ability to work, untethered, regardless of place, time, and institutional or household infrastructure for communication. From our data, the interesting social effects of these affordances are:

- People are more mobile because they communicate anywhere and anytime. For instance, people do not need to remain in their office waiting for an important call.

- People more often use mixed-use settings to make the communications that were previously associated with strong social settings. For instance, work and personal calls are made in cars.

- Because mixed use settings do not have clear, legible cues, more work communications will take place even during "social" time or with family and friends in these settings, and personal communications will take place even during work time or with co-workers in these settings. For instance, people take work calls in a restaurant when eating with friends.

- Further, because wireless technologies are not tethered to specific places, callers often do not know the location of the recipient and cannot moderate their own behaviour according to the norms of behaviour settings. For instance, children call parents at work on the cellular phone.

- As people observe and use wireless technology across more settings and places, the social norms associated with the use of wireless technology in different places fail to become differentiated and clear. (Many organisations such as theatres and fancy restaurants make strict rules or install blocking technology to overcome this trend.) Indeed, the communication norms of even strong social settings begin to become less clear. For instance, some people who would not use their wired phones for this purpose take personal wireless calls in the office.

- To the extent that wireless that technologies are carried on the person, like wallets and purses are, they are considered personal possessions. So, norms governing personal possessions rather than behaviour settings begin to apply to wireless technology. That is, norms of personal discretion and politeness in interpersonal interaction begin to govern the use of wireless technology. For instance, people take personal calls while in a meeting but step out of the meeting to carry on their discussion.

The resulting paradox of these many changes is that, in the previous era, increased mobility led to an increasing separation of work and personal place and life. Wireless technology may be changing this equation. As we become more mobile, enabled by wireless technologies, we use the technology at our discretion. When an employee uses his personal cellular telephone to call his wife from the car on the way to a sales call, is he on work or social time? What if he is calling her from the lobby of his building, or from his office? Does an employer who provides the cellular telephone to his employees have the right to call them during evenings or weekends? Clearly, for the growing ranks of the technology-enabled workforce, wireless technologies make it difficult to draw a distinction between work and social life.

Acknowledgements

We are grateful to those who contributed to this study: from Carnegie Mellon University, Pamela Hinds, Marvin Sirbu, Ben Bennington, Lisa Ann Lightner, Bruce

Taylor, Paul Bistrican, Patrick Keating, Baruch Fischhoff, Jon Peha, Jaime D'Almedia, David Fleming, Jolene Galegher, Arlene Simon, Melissa Wingert; from Bell Atlantic Mobile, Jim McHenry, Steve LaGueux, Jerry Fountain, Tom Callan, John Provinsal, Ben Scott; also to Tina Kiesler (UCLA), Richard Wolff (Bellcore), and Lee Sproull (NYU). Funding for this research was contributed by the Information Networking Institute at CMU, Bell Atlantic Mobile, and an NIMH Research Scientist Development award #MH 00933 to the second author.

References

Barker RG (1968) *Ecological psychology*. Stanford, CA: Stanford University Press.

Cialdini RB, Reno RR and Kallgren CA (1990) A focus theory of normative conduct: recycling the concept of norms to reduce littering in public places. *Journal of Personality and Social Psychology*, 58 (6), pp. 1015–1026.

Gibson B and Werner C (1994) Airport waiting areas as behaviour settings: the role of legibility cues in communicating the setting program. *Journal of Personality and Social Psychology*, 66 (6) pp. 1049–1060.

Hatch MJ (1987) Physical barriers, task characteristics, and interaction activity in research and development firms. *Administrative Science Quarterly*, 32, pp. 387–399.

Hinds PJ (1999) The curse of expertise: the effects of expertise and debiasing methods on prediction of novice performance. *Journal of Experimental Psychology: Applied*, 5, pp. 205–221.

Kiesler S, Gant D and Hinds P (1994) The allure of the wireless. *Technical Report TR 1994-2*. Carnegie Mellon University: Information Networking Institute.

Kiesler S and Sproull L (1987) The social process of technological change in organisations. In S Kiesler and L Sproull (eds). *Computing and change on campus*, pp. 28–40. Cambridge: Cambridge University Press.

Lynch K (1960) *The image of the city*. Cambridge, MA: MIT Press.

Marvin C (1990) *When old technologies were new*. Oxford: Oxford University Press.

Melbin M (1978) Night as frontier. *American Sociological Review*, 43 (1), pp. 3–22.

Norman D (1990) *Design of everyday things*. New York: Basic Books.

Proshansky HM, Fabian AK and Kaminoff R (1983) Place-identity: physical world socialisation of the self. *Journal of Environmental Psychology*, 3, pp. 57–83.

Proshansky HM, Ittelson WH and Rivlin LG (1970) Freedom of choice and behaviour in a physical setting. *Environmental psychology: man and his physical setting*, pp. 177–178. New York: Holt, Rinehart & Winston.

Sproull L and Kiesler S (1986) Reducing social context cues: electronic mail in organisational communication. *Management Science*, 32 (11), pp. 1492–1512.

Stokols D (1990) Instrumental and spiritual views of people–environment relations. *American Psychologist*, 45 (5), pp. 641–646.

Yates, J (1989) *Control through communication: the rise of system in American management*. Baltimore, MD: Johns Hopkins University Press.

Part 4
From Use to Design

Chapter **10**

Welcome to the Wireless World: Problems Using and Understanding Mobile Telephony

Leysia Palen, University of Colorado, Boulder, USA and Marilyn Salzman, US WEST Advanced Technologies

10.1 Introduction

Wireless telephony adoption is on the rise in many parts of the world (Cahners In-Stat Group, 2000), with recent technological innovations continuing to enhance the capabilities of the technology (McGinity, 1999). The market is extremely competitive. Service providers and wireless handset manufacturers, therefore, are faced with the challenge of designing wireless services and handsets to retain customers.

Because wireless technology is becoming a mass-market commodity, new novice, tentative, and price-sensitive customers continue to shift the nature of the user base. In fact, at the time of this study, US service providers typically subsidised the handset prices to lower entry costs, losing money initially with the hope of long-term retention to recoup the outlay and generate profits. With the objective of increasing overall retention, one-year contracts, which often included an initial 30-day grace period, were also common. Thus, the first 30 days were critically important to users and to service providers alike. We found that during this initial period, novice users struggle to understand the technology, how it fits into their lives, and if its benefits exceed its costs. Simultaneously, service providers strive to provide a positive customer experience so that their new users will be retained long-term.

It was within this context that our study was conducted. Our work was initiated by requests by a service provider to investigate handset usability issues. It was important that the handsets they acquired from manufacturers were well designed, because they were subsidising the handsets, and because the handsets were perceived as playing an important part in the early customer experience. Therefore, we set out to explore handset usability in depth, while also taking the opportunity to look beyond the handset at other factors affecting the early customer experience.

This chapter reports on a study of novice users and the problems they faced using and understanding basic wireless telephony service during the first six weeks after adoption. We found that new users typically have poor comprehension of mobile telephony that continues to persist into at least the second month of use (and, we predict, continues well beyond) (Palen *et al.*, 2000). We also found that using service-based technologies like wireless telephony extends beyond a need for users to understand the hardware and software alone. In addition to the hardware and software attributes of the technology, features of the network and marketing communications directly affect how users perceive their handsets to function. In this chapter, we present particular findings about ergonomics, feature use, company communications, and bill design within an overall framework that describes wireless service from a user's perspective. This framework comprises four attributes of the wireless telephony technology system: hardware, software, "netware" and "bizware".

10.2 Related Work

Published studies addressing mobile telephony use and usability issues are only now emerging; some existing work, and other chapters in this book, illustrate why it is important to look beyond the hardware and software of the technology to understand the user experience.

For example, Helyar's chapter in this book (Chapter 13) reports on the results of a usability study of wireless internet service for telephone handsets. Although the study was a traditional laboratory one with emphasis on the handset and services, Helyar found that issues beyond the immediate technology affected its usability. He notes that uncertain costs and billing cycles was a source of apprehension for subjects, and that users were confused by the computer jargon-influenced instructions. However, because this was a laboratory-based study, Helyar was not able to predict how these issues impacted user behaviours and attitudes over time.

Väänänen-Vainio-Mattila and Ruuska (1998, 1999) adapted Beyer and Holtzblatt's (1998) contextual inquiry (CI) techniques to the study of mobile technology. Using this ethnographic approach, they and other Nokia researchers have gathered information about user needs, behaviours and practices by shadowing users in real situations. They have had success in using these techniques to inform the design of mobile technology.

10.3 Method of Study

10.3.1 Our Approach

Our methodological objective was to situate our inquiry in the real activities of novice users, while studying them longitudinally to cover a wide range of naturally occurring user experiences. To this end, we employed three different data collection techniques that we hoped would provide convergent information about users' experiences and their communicative practices.

Our data collection and analysis approaches were in the qualitative tradition of the social sciences, where we tried not to limit our observations by a pre-defined issue space. We attempted to cast a broad net in our inquiries. Additionally, our analysis was grounded in the data, such that the attributes of usability that we describe in this chapter emerged from the data itself.

We conducted multiple in-depth and open interviews over the course of the six weeks immediately following service acquisition. To understand the context of use, we grounded our questions in information that subjects reported in "voice mail diaries", a technique we adapted from a paper-based diary study approach (Rieman, 1993). To tie these observations to frequency of telephone use as a characteristic of communicative practice, we collected data on actual calling behaviour.

As stated, the findings about communicative practice and evolving attitudes about mobile telephony use are reported elsewhere. In this chapter, we describe what we believe to be a set of attributes that address more comprehensive thinking about wireless telephony usability. We do this through illustration of our own findings.

10.3.2 Subject Selection

Nineteen people participated in the study. Subjects were recruited within the few days following their subscription to mobile telephony service. Only about one out of ten people contacted fitted the criteria and was able to participate. Subjects received monetary compensation for their time. One subject, S12, dropped out of the study mid-way, but the partial data for her are used where appropriate.

With one exception, participants were first time users of mobile services; that is, prior to this purchase they had not subscribed to any other mobile telephony service. We did make one exception by including a subject who had previous mobile phone experience, but had special access and safety needs thought to be important to the investigation (Subject 13; see Table 10.1). They were all geographically proximal to the researchers to allow for frequent interviews. Although gender, age, profession, socio-economic status, etc. was not experimentally controlled, our population was varied across these dimensions. Table 10.1 describes in some detail subjects' occupation and lifestyle attributes. A range of professions and domestic situations are represented in this subject pool.

10.3.3 Interview Data

Three interviews were conducted with each subject, lasting each lasting one or two hours. The interviews were open-ended such that central issues were discussed with everyone, but professional and personal factors that were unique to each subject could emerge and be explored and documented. Most interviews took place in our office location, although when possible, some interviews were conducted in people's homes. Family members were invited to participate in the discussion when they were actively involved in some aspect of acquiring, using or paying for the mobile phone.

The first interview was designed to capture the "out of the box experience". This interview was scheduled immediately after the subjects acquired the telephone

Table 10.1 Subject description

Subject	Age (yrs) and gender	Occupation and important lifestyle attributes
S1	46–55 Female	*Artist, doctoral student, and clinical therapist.* Commutes 30 miles to private art studio a couple of times/week, where there is no wireline phone. Shares home with adult room-mates but has own phone line.
S2	16–25 Female	*High School student.* Newly licensed driver. Drives long distances around metro area for sports activities. Shares mobile phone with mother. Lives at home with two parents and older brother.
S3	46–55 Male	*Full-time church pastor.* Office in the home, with a dedicated business line. Works daily outside office in multiple places. Spouse and two teenage children at home.
S4	16–25 Female	*Community college student; part-time retail employee for small shop.* Work and school schedule varies daily. Lives with partner who works regular business hours.
S5	56–65 Female	*Part-time non-profit club manager; church organist; church janitor.* Lives with spouse and one adult son. Uses pager to be on-call as club manager. Church she cleans does not have wireline phone.
S6	26–35 Female	*Homemaker and mother of two; part-time computer system administrator.* Just returned to work. Lives with spouse who works regular work hours, and two toddlers.
S7	46–55 Male	*Construction sub-contractor.* Works on site at multiple locations per week. Spouse stays at home with toddler.
S8	46–55 Male	*Engineer.* Lives with spouse who works regular hours; spouse also has her own mobile phone.
S9	36–45 Female	*Dental assistant.* Lives with spouse who works regular hours; spouse also has his own mobile phone.
S10	36–45 Female	*Meteorologist.* Lives alone. Travels and calls frequently to her large family who lives two hours away.
S11	66–75 Male	*Retired barber; part-time model and law firm courier.* Lives alone. Modelling and courier work is on an on-call basis, which requires immediate attention.
S12	36–45 Female	*Homemaker; student; small-business owner.* Lives with spouse and teenage son.
S13	46–55 Male	*Retired; works frequently outdoors on his large property while in his wheelchair.* Uses phone for safety and accessibility purposes. Lives with life partner; she works part-time out of the home.
S14	16–25 Female	*Professional housekeeper; mother of two young children.* Works multiple locations throughout week and coordinates childcare with her mother. Lives with two toddlers and husband, who works regular hours outside home.
S15	26–35 Female	*Mother of four children under 10; homemaker.* Husband works regular hours outside home.
S16	46–55 Female	*Professor.* Lives alone. Commutes by bus or car 30 miles to work most days of the week. Uses mobile phone in lieu of physical presence in office.
S17	56–65 Male	*Engineer.* Shares a car and mobile phone with wife.
S18	36–45 Female	*Full-time contracts administrator; part-time rodeo teacher; professional rodeo rider.* Lives with room-mate who is a student. Travels to shows, works outside in evening on ranch.
S19	46–55 Male	*Consultant.* Commutes 30 miles to city office by bus. Uses mobile phone to retrieve home office calls in his city office. Lives with spouse and teenage son.

handset, but before they used their phones. Subjects were asked to explore their handsets (and other materials if they desired) as though they were at home doing the activity. Some subjects wanted to consult a friend or family member at a certain point; we documented only the extent to which subjects worked on the phones themselves, following up later on their collaboration with others. Subjects were also free to place phone calls to friends to test their phones, or to customer service to get help. We provided no direct assistance, but did encourage subjects to do what they would normally do to seek help.

The second interview took place approximately two weeks after acquisition and focused on the changes in behaviour and use over that period. At this and the subsequent meeting, changes to the handset settings were noted. The third interview took place after users received their first mobile phone bills, about 4–6 weeks after service acquisition. This interview also focused on changes in behaviour and included a discussion centred on the interpretation of their bills and the calling behaviour documented therein. All interviews were videotaped.

10.3.4 Voice Mail "Diary"

To capture mobile phone activity as well as discoveries and insights that subjects had about their newly acquired phones, we instituted a variation on the "diary" method of data collection (Rieman, 1993). Instead of having subjects record activities on a paper diary, we invited subjects to call in to a dedicated voice mail line and talk about their experiences. They could call in from any phone. In the interest of minimising any bias, we never suggested they use their mobile phone for this purpose. This was an optional activity, but subjects were given $1 for every day they called in, even to report that they did not use their phones that day. Although experimental, this method turned out to be a successful way of capturing activity that is very transient. On average, subjects called in about one out of every two days, although variance was high. This data was also important for reminding subjects during interviews about certain activities that could then be queried further. All diary reports were transcribed verbatim.

10.3.5 Calling Behaviour Data and Phone Bills

Actual calling behaviour data was collected via the network for approximately the first four months of use. The subjects also provided copies of their phone bills.

10.4 Findings: Attributes of Wireless Telephony

Mobile telephony use is affected by four factors of the technology: hardware, software, "netware" and "bizware."

- *Hardware*. In mobile telephony, the handset, battery and charger comprise the hardware component, with the ergonomic issues that all hardware devices face.

- *Software*. Menus and display-based controls comprise the handset's software.
- *Netware*. We refer to netware as the basic mobile telephony service and special services (e.g. advanced versions of call forwarding) that a provider makes available. The type of service – analogue or digital – is also included in this category.
- *Bizware*. Finally, we define bizware as the details of the service agreement, including calling plan, sales policies and customer service.

Below, we highlight some of our findings within this framework.

10.4.1 Hardware

We found that subjects attributed behaviours that originated outside the scope of their telephone handsets to the usability of the handsets themselves. These largely included problems that might emerge as an outcome of poor signal coverage, such as not being able to place a call. Additionally, consequences of business practice, such as policies for assessing minutes of talk time (also known as "minutes of use" or MOU), often conflicted with information independently provided on the handset.

Findings for handset hardware pertained to button size, antenna extension, volume control, battery charger, etc. Of note were the findings about a "dial shuttle" or thumb-wheel on the side of a commonly used handset. Using the dial shuttle to navigate through the feature menus was difficult for new users. The dial shuttle is spun with the thumb of the left hand or a finger on the right hand, moving the cursor up or down through a list of feature menus. It also can be depressed like a button to select a feature in a menu, an action that is not obviously afforded by the shape of the wheel itself. More problematic was that the primary feature menus can only be accessed if the wheel is *first* depressed. The wheel is then spun to move through the list of feature menus. To see the options in a feature menu, the wheel is pushed once again. Therefore, navigating through the feature set requires an alternating series of pushes and spins, with a "push" initiating the navigation.

Trying to push the dial shuttle frequently resulted in unintended menu selections. Furthermore, because of the way the menu structure is laid out, if a user spins the wheel first and then pushes the button, he or she can inadvertently dial a pre-programmed number. One subject, S11, who was very unsure about the operation of the dial shuttle and even more unfamiliar with the computing-related concept of "menu trees", frequently did this, even after six weeks of use.

Because the pushing action of the dial shuttle takes some time for users to discover, many do not discover it without specifically reading the manual or calling customer service. Although providing assistance to users through a customer service function is important and good, it is not a solution for problems that are outcomes of poor design. Customers are frustrated when they cannot figure out basic functionality by themselves, and the service provider suffers from the high costs of telephone support.

Exacerbating this usability problem, we found that novice users do not necessarily know that the software features exist; therefore, they do not even know to look for them! For example, some subjects told us that they inadvertently learned about the dial shuttle operation after calling customer service for another reason: to ask for their mobile phone telephone numbers. (The finding that subjects did not know their new telephone number is a separate "bizware" usability issue and discussed later in the chapter.) Expecting that the customer service representative would refer to records to find their telephone numbers, the representative instead told them that it was available as a software feature on their phones. The customer service representative walked these subjects through the steps of finding their phone numbers, which included pushing and spinning the dial shuttle. It was at this time that these users not only learned their phone number but also discovered the many other features available on their handsets as well.

Another problem subjects encountered was discovering how to change the earpiece volume, which was controlled directly by a dial shuttle or buttons on the side of the phone. Earpiece volume also could be changed via software menus; however, controlling volume this way required users to interrupt their call, look at the screen, and navigate the menus. Consequently, subjects often suffered through phone calls set at inadequate volumes because they did not know how to easily change the volume *during* a call. Furthermore, this problem was not isolated to single incidents but instead spanned long periods of time. Subjects kept inadequate volume level settings for days or weeks because they did not know how to change them.

10.4.2 Software

We define software as features on the phone that are accessed via menus displayed on a small screen. Our interviews (including inspections of subjects' phone settings) and diary reports revealed which software features were used by the most users; they also enabled us to determine roughly when users first used the features. We observed how situated actions lead to the discovery of features, and how this discovery of (or lack thereof) affected subjects' experiences with the phone. We also saw that some potentially helpful features failed to be used simply because of the location in the menu structure. We discuss examples of these here.

Most commonly used features included the save/program numbers feature, ringer volume, and earpiece volume, although earpiece volume took some time to discover. From a service provider's perspective, discovery and ease of use of all of these features are critical, as they help users to feel comfortable with their phone service and to use it effectively.

The save/program numbers feature, which allows users to store telephone numbers in the handset phone book, is an example of a feature that was discovered and valued by many users early on. It was used by many subjects (74%) by, or during, the first meeting (immediately after acquisition) and by 89% of the subjects by the end of the study, leaving 11% (or two subjects) who never discovered the feature. In some cases, discovery was a result of exploration: On the most commonly used phone, the phone book appears when users spin the dial shuttle, making it easy to find. In other cases, the phone book's discovery was a direct result of calls to customer service concerning how to use the phone or service. In either case, users who discovered this feature reported that it was particularly useful for saving numbers they might otherwise have forgotten, as well as for placing calls quickly.

Another feature, ringer volume – which controls how loud the phone rings – was used by 37% of the subjects by the first interview, 74% by the second interview, and 84% by the third interview. Although an important feature for people to know about early on, the ringer volume controls are modal and hidden deep within the menus, requiring users to discover this feature through trial and error. In fact, several subjects reported searching for this feature immediately following or in anticipation of the phone ringing in a publicly inappropriate situation. Interview data and usage patterns also suggest that, after discovering this feature, users grew to rely on it to support the usage of the phone in a variety of contexts with different ambient noise levels and in different social settings.

Earpiece volume was another feature that needed to be salient early on, but was only gradually discovered by our subjects. Use after purchase was low at 11%, and only slightly higher (16%) at two to three weeks. At the final interview, about six weeks after acquisition, only 47% of the subjects had discovered it. Many of the remaining subjects complained about volume level without knowing it was adjustable.

Backlight is the feature that controls the illumination of the small telephone screen in low light conditions, which users encountered when using their phones in a variety of real world situations. Thirty-seven per cent of our subjects increased the duration of the backlight. They found that when using their phones at night or in dimly lit places, the default time was not long enough. Furthermore, it was the very experience of using their phones in low light that led most to discover the existence of the adjustable backlight feature.

In addition to learning which features were used by subjects, our investigation also clarified how some features were used in relation to other information provided by the service provider. For example, we found that subjects thought that the minute counter was initially helpful, but later discovered that the service provider's tally was in conflict with the handset readout. In this case, the handsets' minute counters used different rules than the provider for tallying minutes of use, and the handsets did not provide subtotals that were meaningful within the context of common service plans (e.g. subtotaling minutes based on time of day or weekday vs. weekend). Consequently, when subjects called an automated service that reports number of minutes used, some were quite concerned that the number was different from what was reported on their handsets. Furthermore, subjects tended to rely on the information from their own handsets foremost and expressed a growing sense of mistrust of the service provider's numbers as a result of this conflicting information.

10.4.3 Netware

Netware encompasses the type of mobile telephony service (digital or analogue); special calling plan services like call waiting, call forwarding, and other provider-specific services; and calling area coverage and coverage consequences on call placement.

10.4.3.1 Type of Service

We found that the difference between analogue and digital service was unclear to most subjects; in fact, most did not know what kind of service they had. All sub jects in this study had *only* digital service (the only option with their service provider). When subjects owned dual mode phones (phones that can switch between digital and analogue service), they seemed to be further confused by the meaning of analogue and digital signal.

A dual mode phone will switch to analogue mode when digital signal is weak, even if a caller is in his or her home area. Because subjects did not know that their service provider dealt only in digital service, they did not then realise that whenever they were in analogue mode, they were automatically "roaming" and incurring additional charges. (The concept of roaming, which was a source of significant confusion for subjects, is discussed later in this section.) Consequences to this usability issue were unexpected charges, which led to additional calls to customer service for explanation, according to subjects' self-reports.

10.4.3.2 Service Coverage Areas

We found that most subjects did not understand the factors that affect service coverage and call quality. Specifically, new users did not understand the relationship between digital coverage quality and these factors:

- geographical terrain
- calling inside a building versus outside
- building composition materials
- calling near a window when inside
- call load or traffic as it is affected by time of day
- use of phone antenna.

Furthermore, several subjects did not see that there were signal indicators on their handset screens, and did not know how to interpret them.

A poor understanding of the natural limitations of wireless telephony service led subjects to incorrectly attribute problems to the service provider's quality of service. Subjects also blamed their handsets for operating poorly, wondering if a low battery limited reception or if their handset was otherwise defective. For example, S4 (who has a single digital mode phone that cannot switch to analogue) describes her experiences when trying to place a call in a building:

> I was [at school], and my phone kept cutting out and it kept saying that ah- the call failed, and then it said that the call was lost. So I am a little bit frustrated with that because I thought it would be, ya know – I've seen other people use phones in a

building and theirs works fine and mine just cut out. And so I was a little mad that I had to go outside of the building to use it ... (S4)

What S4 did not recognise was that other successful calls she witnessed may have been made on analogue-supported phones; or perhaps the calls were placed when near a window, or at a time of day when calling traffic was low. In the next example, S6 understands that signal quality is affecting her placing calls, but still wonders if this is somehow related to the operation of her handset:

I was using my phone today but I had quite a bit of trouble. I didn't get any calls today and when I got home I had six messages on the machine, and it was in analogue roam all day no matter where I was. No, actually that's not true. When I left the school that I was working at I'd be in regular mode, but at the school I was in analogue mode, which is ridiculous: [these are local schools] ... So I am definitely having a problem with the phone. Trying to figure out you know why it's doing that ... (S6)

These particular usability problems can affect customers' confidence in the operation of their handsets as well as the quality of the service provided by the phone company.

10.4.3.3 Special Network Services

Wireless telephony service providers can offer a variety of special network services, including call routing, voice messaging and data services. These special services can be used to encourage customer loyalty, to increase revenue and to attract new customers.

Our research revealed that appropriate service assignment is highly important for the success of network services. However, we also found that striving for good matches between customers and services can be at odds with sales goals. In many business organisations, salespeople are typically rewarded by their volume of sales and are given additional incentives for assigning special services. Furthermore, because the sales function is often out-sourced to a third party business, salespeople fail to feel the downstream effects of an inadequate sales interaction. From a salesperson's perspective, there is little incentive for taking the time to develop a deep understanding of a customer's needs or to explain ways to effectively manage the service. Indeed, they may not even know enough about the service to assign it to appropriate customers.

An example that illustrates these challenges is call routing service sales. Call routing service allows all calls placed to one's home or office number to be routed to one's cell phone instead. This enables users to distribute only one telephone number and receive their calls anywhere. In general, call routing services between wireless and wireline telephones are highly successful services for those for whom the services are a good match. In our study, many subjects found these services extraordinarily convenient because they allowed them to receive their calls anywhere, without having to wait at a wireline phone. These services also freed them from having to manage which phone number (home, work, mobile or both) to distribute.

However, some subjects signed up for these services not fully understanding them. Thus, they neither understood which version of the call routing service was best suited for their life-styles and communication needs, nor even how to use the service. As a result, these users experienced difficulty in managing communications on the road, at work or at home. For example, one user (S7) wanted his calls routed to his wireless phone only when his wife was not at home to pick up the calls. Unfortunately, he had not selected the call routing service best suited to his needs and he did not understand how to manage the service he had selected (e.g. turning it on and off). When S7 found that he was receiving his wife's calls while he was at work, and that she could not receive calls at home, he decided to keep his wireless phone off most of the time to keep the calls from being re-routed. This caused him to question the value of having a cell phone at all:

> The other thing that I'm still trying to get done, and I am having a lot of difficulty is getting the service that hooks my – connect my – routes my home phone calls to my cell phone [turned off]. … It's really a drag, because I have to have my phone turned off, if [my wife] is at home and wants to get phone calls, which really kind of – what's the word? – cancels the points of having a cell phone. (S7)

Determining the characteristics for good service assignment and carefully explaining that service to customers, therefore, are clearly important usability challenges for service providers. This failure to communicate with customers at time of sale means more calls to customer service downstream to either seek an explanation of the service, or to have it disabled or removed. It can also mean that users fail to use their mobile phones as they otherwise might, especially if they have to habitually keep their phones turned off to avoid problems with their special services.

10.4.3.4 Long Distance and Roaming

Although long distance is not a new concept to people, new users are confused about when various charges apply for long distance and roaming. For example, about one-third of our subjects incorrectly thought they were charged for incoming long distance calls. Because new users are uncertain of cost structures and unsure of long distance operation, they tend to err on the side of conservatism by limiting long distance phone calls to prevent unexpected costs, a finding consistent with Helyar (Chapter 13). The bill is often the first time the relationship between their calling behaviour and charges is clarified for users, which happens about one month after acquisition.

In particular, we found that the concept of "roaming" was difficult for new users to comprehend, and was often confused with long distance calls. Roaming occurs when a person is outside their "home" coverage area and makes or receives a call that a different service provider local to that area handles. Many subjects had very poor understanding of this; in fact, their comprehension was so confused as to be indecipherable by the researchers at times! In most cases of confusion, subjects understood roaming to be a function of the destination or origination of phone calls, not a function of the location of the mobile phone and the user. So, a call to or from an area outside one's home area would be considered a "roaming" call, hence the confusion with long distance. Other subjects erroneously thought that "roaming" was the state of the phone "looking for signal", and so was not associated with any potential calling charges. One unfortunate consequence of this interpretation could mean a surprisingly high phone bill for the unsuspecting user.

The following examples show how subjects confused long distance and roaming, and how other aspects of service were also sometimes confused as a result. In the first example, S1 was probably trying to put her dual mode phone into digital mode only to avoid roaming charges when she was in analogue-only coverage areas. In addition to demonstrating a poor understanding about how coverage works (by incorrectly referring to non-existent "satellites" and "roam service"), she also confuses a free long distance benefit of the service provider – toll-free in-state calls – with roaming. Although roaming would prevent S1 from making toll-free in-state calls, she incorrectly equates long distance calling with roaming:

> I called customer service and learned how to put my phone on only picking up your satellite so that I could call anywhere in the state for free. And then I came up to [a town in the mountains] today and I couldn't get any service so I went back in under "Network" [features] and figured out how to put my phone back on analogue roam.... I like having this option and when I [come] back down [from the mountains], I will probably put my phone back on, just having [the phone company's] roam service. (S1)

In the next two examples, both S8 and S18 also equate long distance and roaming:

> (A single digital mode subscriber) I tried to make a long distance call out in Los Angeles to see if I could get through to [my wife at home in Colorado] and it wouldn't work, which I don't know if that's a digital thing or a [service provider company] thing. But it's – I'm getting more anxious to turn this thing over to a digital/analogue so that you can hopefully get a roaming connection long distance with your cell phone otherwise it's kind of worthless. (S8)

S18 demonstrates her confusion between long distance and roaming by using these concepts interchangeably:

> Catching you up on all the goings on of the weekend: I was out of town again for the weekend, so I didn't call you because you know those *big* roaming charges that scare me to death.... Sorry to report four days in one day, but I was out of town and those good long distance charges just, you know, make me nervous. (S18)

In addition, one-third of subjects thought that they might be charged for incoming long distance calls. For example:

> I am actually kind of wondering whether I like to answer long distance calls or not. Because if I am paying for the long distance call even though they are calling me, [I'd prefer to use] my home phone. So I don't know. Maybe that's not true. Maybe I am not paying for it, they're paying for it, like it's supposed to be. But anyway, that would be a good question probably for the phone company ... (S6)

The effects of these misunderstandings on mobile telephone use can be considerable if left unchecked, with new users hesitant to make and receive certain calls in fear of incurring high charges. In some cases, calls do actually incur higher charges (when roaming, for example), but without users feeling certain about these conditions, they cannot fully trust the usability of their phones. In summary, although users are familiar with the concept of long distance under the wireline model, they confuse it with the concept of roaming under the wireless model, and therefore do not understand how and when they will incur long distance charges.

10.4.4 Bizware

Bizware refers to the non-technical components of mobile telephony, including calling plan agreements, service provider-specific business policies, sales communications, customer service communications, marketing promotions communications, manuals and other information resources, and phone bills. Bizware quickly emerged as a prominent factor in the usability of wireless telephony for our subjects. Bizware issues – most of which we did not anticipate – emerged from our grounded, longitudinal approach to the investigation.

10.4.4.1 Calling Plan Agreements

Agreement plans specify how much "airtime" (in minutes) one has to use over the course of typically a month. Calling plans vary within and across providers: different levels of airtime minutes are available, and airtime can be distributed across "peak" and "non-peak" times.

For novice users, deciding what calling plan to buy can be a challenge. Plans are usually arranged by total number of airtime minutes per month: the greater the minutes, the greater the price. There are three buying strategies users employ: selecting by price, selecting by amount of airtime minutes, or finding some balance between the price and amount of minutes. All but three of our subjects opted for the lowest price plan. Most of these subjects appeared to select on the basis of price. As novices, they seemed to want to gingerly enter the market without incurring too many costs.

As a consequence of selecting by cost, users constrain their number of airtime minutes. This, in turn, means that they must fit their communicative practice to these imposed limitations (or incur much higher costs for surpassing the limit). In contrast, selecting by minute plan allows users to fit the technology to their communicative practice. However, even though the latter arrangement would seem desirable, subjects reported some difficulty in anticipating their calling time in terms of minutes over a month. Although people may be accustomed to managing their long distance calls by amount of minutes, anticipating one's own calling behaviour over the span of a month with a new technology is not familiar.

Subjects who selected by cost were concerned about exceeding the number of minutes in their plans; some over-constrained their calls and used far less than their plans allowed. Only two subjects who selected by price surpassed the calling minute limitation. The three subjects who selected their plans by number of minutes (and who did not select the minimum price plan) used far fewer minutes than their plans allowed.

Tracking customers for the first cycle of service yielded this information about calling plans and early behaviour. Failure to target a suitable calling plan is a usability issue that affects the customer experience, and can in turn affect a customer's relationship with the business provider. We hypothesise that because novices are uncertain about how their actual calling behaviour will match with their calling plans, they initially constrain their calling to avoid expensive surcharges. This might mean that customers do not explore uses of their phones that they might otherwise. Furthermore, customers may settle into this pattern for the long-term.

10.4.4.2 Provider-Specific Calling Plan Policies

Provider-specific calling plan policies yield additional information that users must assimilate and incorporate into their calling behaviour. Our investigation revealed that if these policies were not understood during the sales transaction, there was little opportunity for subjects to learn about them later.

For example, "peak" and "non-peak" calling times are decided by the service provider, and might vary from conventional wireline peak times. In particular, the subjects' service provider's non-peak times were restricted to weekends. The service provider actually defined weekends by the clock and calendar: midnight Friday to midnight Sunday. However, subjects did not know about this policy. Some assumed that the weekends would follow actual clock and calendar time, but others used their experiences from wireline telephony to assume that weekends started at 5pm (or 7pm) on Fridays and ended some time on Sunday or even Monday. Because non-peak minutes are often given generously in service plans, users could potentially use the more precious peak time minutes during a time they incorrectly thought was non-peak.

Other policies could benefit users, but it is a problem when the users do not know about them. For example, one-third of the subjects did not know that all in-state calls were toll-free. Additionally, one-third did not know that the first minute for incoming calls was not charged against their minutes of use, and half of the subjects did not understand the benefits of this policy – they won't be penalised for wrong numbers or other nuisance calls. Even if a service provider is doing a reasonable job in conveying information to most of their subscribers, novices are a special population that needs additional assistance to assimilate this variety of new information that differs from older models of wireline telephony.

10.4.4.3 Promotions and Marketing

Special promotions by marketing divisions create expectations for users that might affect their phone use. For example, the subjects' service provider had offered a special promotion to them by offering free weekend minutes. In this promotion, the offer meant that substantial non-peak airtime minutes were included as part of the plan. However, the meaning of "free" was not clear to all subjects, resulting in significant impacts on calling behaviour. For example, one of our subjects erroneously thought that his calls were free of charges on the weekend, specifically long-distance charges (which were not free). Acting on this, he saved calls to his brother (who lived out-of-state) for the weekend, deliberately using his wireless phone rather than his wireline phone. Upon receipt of his phone bill, he discovered the long distance charges.

Why might this have happened? Further investigation of this issue revealed that there were three different connotations of the word "free" communicated by the service provider through promotions or in the literature. In addition to "free weekend minutes," there was "toll-free in-state" calling, and "first incoming minute free".

Toll-free calls meant that long distance charges are not incurred for in-state calling. It is this meaning of "free" that subjects seem most inclined to adopt and apply to other instances of "free". Note that airtime minutes were still docked against the calling plan for in-state toll-free calls. In contrast, "first incoming minute free" meant that the first minute of an incoming call was not docked against calling plan minutes. Some subjects thought this special policy meant that the first minute was free of long distance charges. It was, in fact, free of long distance charges, but that was because there were no long distance charges for incoming calls! Some subjects thought that the "first incoming minute free" therefore implied that the *other* minutes of an incoming call were not free, that they incur additional charges. This confusion was compounded by the belief held by some subjects that they pay for incoming long distance calls. Instead of perceiving this service as a benefit, some subjects assumed the worst.

This example demonstrates how promotional language can affect a user's relationship with their phone. Although promotions are usually the responsibility of marketing, their decisions can directly affect the usability of the technology they are trying to promote.

10.4.4.4 Sales and Customer Service

Sales and customer service desks are also often distinct organisationally. Salespeople are typically motivated to sell by volume. Often there is little time to adequately ascertain new users' comprehension of their new service. Furthermore, customers may in fact be told much of the information they need, but novices do not know what they need to know and retain.

We found that new users often are satisfied by the amount of information garnered during the sales call, only to learn how little they really understand after using their phone for a couple of days. Consequently, customer service received many inquiries about wireless telephony basics from our subjects. In the initial days after acquisition, subjects seemed to need explanations about how to use the special services they received, as well as confirmation of their wireless telephone numbers. Sometimes subjects would call when their phones behaved in an unexpected way. The following case is an example of how information received in the sales call was either not adequately conveyed or retained by customers. S5 and her husband did not remember receiving the special call routing service. They had received it as a free offer but, like other subjects, did not remember the details from the sales call. In fact, because the service was free, we believe that some subjects may have quickly agreed to it without understanding its implications. Because S5 did not have a name to put to her phone's behaviour, she could not troubleshoot the problems she encountered by looking through the manual herself:

> My cell phone just picked up a call that came in, I mean, it came in for my home phone *again* just like the other day. I looked through the book, I can't figure out what's going on. I've tried ... to stop any call forwarding. I am going to keep looking but ah you know, I am really confused why it just comes right through from ringing on my phone from the house phone over to the cell phone. (S5)

10.4.4.5 Order Confirmation

Our investigation included an evaluation of all the paperwork that came with subjects' handsets, including the purchase order. All subjects had their handsets delivered by mail. Because of the great deal of information imparted at the initial sales call, novice users need immediate confirmation of their order. However, our subjects found that certain details of order confirmation (like calling plan minutes and services ordered) did not come until the first phone bill, which arrives about one month after acquisition. Similarly, users needed a written reminder of their telephone numbers and did not think they received one. The number was, in fact, on the purchase order that came with the handset, but was listed as a string of numbers without dashes or spaces (e.g. 3035551212) and labelled as "mobile ID number," phone company nomenclature that means little to new users. Users who did not know about a software feature on their handsets that listed the unit's phone number had to call customer service to get their numbers.

10.4.4.6 Manual and Supplementary Materials

Our study also included an evaluation of how the manual and other supplementary materials were used. Mobile telephony has its own terminology that new users must learn, with much of this language appearing to the user for the first time in the manual and accompanying materials. Some of these terms are inherited computer lingo, which assumes computer experience. However, there were subjects even in our small subject pool who did not have a computer background. Words like "scroll", "icon" and "select" have found their way into manuals, for example, and were a source of confusion for some of our subjects. Helyar (Chapter 13) reports similar findings. Other new terms like "analogue", "digital", "roaming", "airtime", etc. are often used to describe new and tricky concepts as though they are self-evident.

10.4.4.7 Phone Bill

The telephone bill is an important part for understanding wireless telephony, especially the first bill. Our subjects perceived the bill as their first opportunity to compare their calling behaviour to real costs incurred, and to clarify and confirm calling plan details.

Our investigation revealed that formatting left some important information unclear to our subjects, which in turn led to follow-up calls to customer service. In particular, terminology and formatting masked actual calling activity, particularly with respect to weekend versus weekday calling. These problems also misled customers into thinking they were overcharged when they were not. Finally, the bill – which subjects expected to explain their orders in absence of earlier confirmation – did not communicate calling plan details comprehensibly.

We also found that subjects who purchased service under a promotional deal expected to find those promotional names clearly indicated in their bills; they were concerned when they couldn't easily identify those special deals. User struggles here clearly demonstrated that promotions created by marketing divisions also have to be reflected in customer bills. Socio-organisationally, however, this can be a challenge for service providers when these departments operate independently.

When novices acquire wireless telephony service, they have many questions and misconceptions. We found in this particular case that these concerns were not addressed during the sales process, and were even exacerbated by the introduction of new concepts and choices in service. During the first month of use, customers are in a learning period during which they encounter particular problems that we have summarised here. In response, some customers call customer service, while others wait for the first bill for confirmation and clarification of their orders, and to help them understand their own calling behaviour. Therefore, we found that an easily understandable bill is critical for customers not only for its standard financial purpose, but also from the perspective of crystallising users' fledgling mental models of wireless telephony service. When the phone bill fails them, we found that users will call in to customer service, often after spending a good amount of time trying to decipher their bills. Even worse, incomprehensible bills fuelled suspicion in several subjects that they were being over charged by their service provider.

10.5 Conclusions

User comprehension of mobile telephony requires understanding of new attributes not required of other stationary, non-service-based technologies. In addition, mobile telephony is still new, with all the imperfections in the conceptual use model that any new technology faces. Therefore, mobile telephony requires assessment beyond the laboratory. Had this investigation had been restricted to a one-time, laboratory investigation, our findings would have centred on the handset, with little to no understanding of the other areas of concern outlined here. Among other limitations, we would have not understood the sometimes complex interaction between the handset and service, and how this created a different class of comprehension problems. In addition, the importance of other factors such as provider plans, phone bills and promotional language in comparison to handset design also would have been masked.

The four attributes of wireless telephony outlined here – hardware, software, netware, and bizware – each must be understood by users to ensure proficient use. In enumerating these attributes, design improvements become much clearer. We describe some of these opportunities here.

10.5.1 Hardware

Even with a limited sample of handsets, it was clear that hardware issues can result in significant problems. Controls that are hard to operate or controls that are accidentally activated can affect the user's satisfaction with the product, drive up calls to customer service, and ultimately result in lost revenue to the provider. Thus it is important that service providers evaluate the usability of the hardware they purchase from handset manufacturers. Furthermore, because users tend to misattribute handset problems to problems with their service and vice versa, usability data based on customer's verbal complaints can be misleading.

10.5.2 Software

Our study illustrated how basic features such as ringer and earpiece volume can be hidden within and by the software interface, preventing their discovery. As with the hardware, it is clear that this negatively affected the user experience and could ultimately cost service providers. Making critical features salient is likely to ease difficulties of early user experience, reduce the learning curve, and help users integrate the technology into their daily lives.

10.5.3 Netware

One apparent problem with netware is that it is hidden to the user. Another is that it often reflects the idiosyncrasies of the technology supporting it. The misconceptions users have about various aspects of netware are many; simplifying the conceptual model is of critical importance. Interestingly, some effective solutions can come from modifying other attributes of the technology such that netware functionality becomes "invisible" from the user's perspective. For example, flat rate programs can eliminate the need to even worry about the underlying network infrastructure that often drives roaming and long distance charges.

10.5.4 Bizware

Confusion from bizware can stem innocently from historical practices and organisational structure of the service provider. Responsibilities that seem singular from the user perspective (sales and customer service) may in fact be distributed across the organisation with little understanding of how one unit's activities affect another downstream. Therefore, addressing bizware usability improvements can be quite complex from the service provider's point of view, but could reap tremendous benefits in reduced costs and increased revenues.

A good fit between customer needs and type of service is also critical, and is an outcome of bizware practices. In our study, each customer had particular social, mobile and communication needs that were the key drivers in how he or she realised the benefits of the phone, which we discuss at length in Palen *et al.* (2000). When service plans were not suited to these needs, users suffered. Thus, we believe that achieving a good match between the user needs and the wireless services can go a long way to ensuring that new customers will make it through the early trial period, as well as achieve communicative practices that resonate with their lifestyles. When users are at odds with their phone or, worse, are afraid of incurring high costs because they do not understand fee schedules, they under-constrain use.

We have outlined four attributes of wireless telephony that we believe articulate the sources of user confusion with the technology. It is our hope that such a framework helps manufacturers and service providers to parse and categorise usability problems, as well as to understand cause-and-effect relationships between different attributes. We believe that this kind of information is critical in the development of strategies for offering *successful* wireless services, services that are useful and enjoyable for users and profitable for the wireless industry.

References

Beyer H and Holtzblatt K (1998). *Contextual design: defining customer-centered systems*. San Francisco, CA: Morgan Kaufmann.

Cahners In-Stat Group (2000) *Cellular Market Goes Ballistic, '00 Subscriber Forecast*. Report GW00-04SU.

McGinity M (1999) Staying Connected: Flying Wireless, with a Net. *Communications of the ACM*, 42 (12), pp. 19–21.

Nippert-Eng C (1996) *Home and work: negotiating boundaries through everyday life*. Chicago, IL: University of Chicago Press.

Palen L, Salzman M and Youngs E (2000). Going wireless: behaviour and practice of new mobile phone users. *Proceedings of the ACM CSCW 2000 Conference on Computer Supported Cooperative Work*, Philadelphia, PA, pp 201–210.

Rieman J (1993) The diary study: a workplace-oriented research tool to guide laboratory efforts collecting user-information for system design. *Proceedings of the ACM INTERCHI '93 Conference on Human Factors in Computing Systems*, pp. 321–326.

Väänänen-Vainio-Mattila K and Ruuska S (1998) User needs for mobile communication devices: Requirements gathering and analysis through contextual inquiry. In C Johnson (ed). *Proceedings of the First Workshop on HCI for Mobile Devices*, GIST Technical Report G98-1, Department of Computing Science, University of Glasgow, Scotland, pp. 113–120.

Väänänen-Vainio-Mattila K and Ruuska S (1999) Designing mobile phones and communications at Nokia. In E Bergman (ed.). *Information appliances and beyond: interaction design for consumer product*. San Francisco, CA: Morgan Kaufmann.

Chapter **11**

Framing Mobile Collaborations and Mobile Technologies

Elizabeth F Churchill, FX Palo Alto Laboratory Inc, USA and
Nina Wakeford, Department of Sociology, University of Surrey, England

11.1 Introduction

Recent years have seen a marked increase in the production and promotion of portable, wireless communication devices: mobile phones with internet access, wireless PDAs such as the Palm VII and smart pagers such as RIM's 850 and 950. Some claim the presence of such devices in the hands, bags and pockets of so many people heralds a new world of work in which people can be reached and information accessed "anywhere, anytime". Whether or not access to information in itself can promote new working practices, individuals whose lives revolve around movement between work sites have been singled out as an obvious market for such portable wireless communication devices. Using these devices such "mobile workers" can be in touch with colleagues, collaborators and clients "24/7", and still sustain non-work social relationships due, apparently, to their constant connectedness whilst mobile.

In this chapter we have two goals. The first is to address the design of mobile technologies. The second is to illustrate our design approach, wherein we consider local practices of technology use, but also the broader cultural context in which technologies are designed, produced, bought, sold, used and redesigned. Our ultimate design aim is to build on existing practices, but also to consider possibilities for the development of innovative technologies that enable new, complementary, practices.

11.1.1 Addressing Mobile Work and Mobile Technologies

In our analysis of mobile work and mobile technologies we have considered commercial representations of mobility, of mobile work and of mobile technologies, and compared their imagery with actual practices of using such technologies. Our aim is not simply to render visible the divergence between common stereotypes of mobile work, as embodied in marketing/advertising rhetoric, and actual use.

Rather, we position both acts of representation and consumption within the "circuit of culture", a model which has been used to examine the emergence of the Sony Walkman, an earlier mobile technological device (du Gay *et al.*, 1997). Taking this model as our inspiration, we wish to argue that the consumption of personal technologies cannot be separated from the way in which technologies are represented in the broader culture, just as an analysis of the consumption of enterprise software cannot be separated from an understanding of the organisation within which it is installed and used. Therefore, we wish to argue for the explicit consideration of the *representation* as well as the *consumption* of device types, "families" or genres within the design/production process.

To argue for the inclusion of analyses of consumption as practice in design is not new (see for example, Ehn, 1988; Luff *et al.*, 2000; Suchman, 1987; Simonsen and Kensing, 1997). However, while the argument for analyses of consumption is well trodden, the argument for analyses of representation is less familiar. After all, representation is often seen as only tenuously related to actual practice.

However, we suggest in this chapter that representation plays an important role in the production as well as the consumption of devices. How does this happen? Du Gay *et al.*, note the design/production of artefacts often involves recourse to imaginary consumers of potential technology. Our own experience confirms this observation. Sometimes reference to these imagined characters remains implicit. When explicitly part of the design process this is often referred to as "persona-based design", imaginary characters also appear in "scenario based design" (e.g. Carroll, 1995; Erickson, 1995). Unfortunately, these imaginary users are too often fleshed out from unsystematic, partial knowledge and common stereotypes. In our experience, they often lead to incremental feature design with a flavour of "Wouldn't it be neat if it (the technology) did X?". We argue that such stereotypes are, in part, derived from the idealised representations that can be seen on billboards, in magazines and in television commercials. The process of assimilation is complex and indirect, but the fact that we build informal, idealised models of activity is exemplified in moments when stereotypes are challenged and *we feel surprised*.

We therefore argue that an analysis of common representations in the form of advertisements can yield potential insights and clues as to how artefacts are being *viewed*; that is, how their acquisition is being (has been) motivated and what standards are held, against which their "performance" is evaluated. Such insights can feed iteratively into the fieldwork and design process where models of idealised activity can be explicitly questioned.

11.1.2 An Outline of Our Design Approach

We have been using a three-pronged design method as part of the production process of artefacts. The three activities are:

1. Content analysis of common representations of artefacts to determine key concepts and fundamental narratives that drive interpretation and consumption within certain areas (e.g. mobile work, business practices). We have focused our attention in this chapter on advertising as a form of representation that adds to

common discourses of mobility. We believe that advertising plays a strong role in discourses of mobility, in part because so many of the technologies are new and do not have a long history of practice-based stories of use. Of course, there are many intermediate factors and in no sense are we suggesting a direct link from advertising rhetoric, to discourses of mobility, to design. Rather, we are suggesting a subtle and indirect link, and suggesting analysis of common representations of devices is useful as an analytic tool when carrying out fieldwork, by supporting the formulation of critical questions that offer a platform from which to elicit insights and view existing practices.

2. Field work analyses of practices of consumption around devices. We propose careful engagement with the uses of devices and consideration of anticipated (Why did you buy it? What did you think it was going to do for you?) and actual practices including unexpected uses, which may of course not be explicitly articulated but can be seen in field work observations (What do you do with it? What works? What doesn't work? Was it what you expected? Did it live up to expectations?).

3. Transformation of field work observations into frameworks that describe or summarise activities, and within which related technology designs may be elaborated. Such frameworks provide a grounding for the creation of new scenarios of use and, from these, the production of novel designs. We believe such frameworks are essential as an intermediate representation between fieldwork and design. Like a diagram, they can capture a thousand words and yet remind us of issues that may be smoothed over.

We illustrate this process as we have applied it to the analysis of mobile work and the design of mobile technologies and mobile working practices; we suggest two dimensions of social experience around which design scenarios can be elaborated. Before illustrating our process, we describe du Gay *et al.*'s circuit of culture in more detail.

11.2 Creating the Consumption of Mobile Devices

Perhaps the most common cultural icon of the portable technological device before the mobile phone was the personal stereo. Although personal stereos might seem to be far removed from mobile office applications of the current generation of information and communication devices, they have played an important part in introducing the mobile device to everyday life. The original model of the Sony Walkman was produced in 1979. Two features of the Walkman are particularly important in the process of production, representation and acceptance. Firstly, the Walkman represented an engineering feat of miniaturisation; reducing the size of equipment needed to listen to pre-recorded (or user compiled) sounds. Second, it represented a shift of a private activity into public space. The user could now transport an individualised soundscape which might replace or overlay that which was already occurring in a public space. Other mobile devices have similarly been designed to fit into these logics: work tasks which might previously have been tied to one geographical space in the office, at a desk or on a computer network, may now be carried on via the mobile phone or pager.

11.2.1 The Circuit of Culture

In their cultural history of the Sony Walkman, du Gay and his colleagues indicate that such features of the personal stereo can be understood by analysing how the Walkman is positioned within several social processes: representation, consumption, production, regulation and identity. By interrelating these processes du Gay *et al.* map out a complex model (Figure 11.1). Crucially, this model stresses the transformation and negotiations which may happen within as well as between social processes. For example, du Gay *et al.* stress that designers are the cultural intermediaries[1] who influence, both physically and symbolically, the form of devices. Yet there can be an ongoing tension between the cultures of production of designers and the cultures of consumption of users. In the early models of the Walkman, the designers had built in two headphone sockets for joint listening. This technical feature emerged from ideas about social practices of listening which were not borne out by actual use. Soon the second headphone socket disappeared from all models, although clearly its technical functionality was never in question.

A further examination of each of these processes suggests a rich context for the development of frameworks for the study and design of mobile devices for collaborative work.

11.2.1.1 Representation

Objects acquire meaning through practices of representation. Often the most striking of these practices occurs through the discourses of advertising (see also Sivulka, 1998). Amongst the many tactics which advertisers have used to represent the Walkman have been the associations with youth, individuality and fashion. Even more strikingly there has been a continued emphasis on the portability of the device in terms of size and weight. The Walkman is particularly significant, as it has become the measure for emerging models of other portable devices. Early

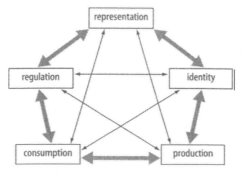

Figure 11.1 The circuit of culture

adverts for the CM-R111 mobile phone produced by Sony actually pitch the mobile phone directly against the Walkman using the slogan "Phone your friends and tell them how big your Walkman is".

Advertisements for the Walkman also stressed the capability of one technological device to serve many different populations. By using pictures of shoes, mocked up as if they were heads each wearing a Walkman, Sony played into this idea of highly flexible customisation. Customisable faceplates and ring tones are of similar appeal in the mobile phone market.

11.2.1.2 Identity

Representations tend to ascribe to users specific social identities. In the case of the Walkman, youth cultures were signalled through iconography and the association of the Walkman with particular kinds of music. In the early days of pagers "on call" workers were represented as the users. These tended to be professionals such as medics or lawyers. Currently mobile devices such as mobile phones and pagers are represented both as potential work tools and as objects which permit (and may even consolidate) particular kinds of social networks in leisure time. Whereas the Walkman was primarily a device which transformed the locatedness of a leisure activity (listening to music), other mobile devices can be used both for "official" collaborative work processes and "unofficial" social ends. To a large extent it is the apparent compatibility of performing specific but varying activities through the same device that is at the core of the ways identities are invoked in the representations of mobile devices for work practices. This is apparent in the advertising images described in the next section.

11.2.1.3 Production

Advertising encourages one particular culture of representation. Cultures of production are also significant in the cultural biography of the object. Du Gay *et al.* point out the ways in which the designers at the Sony Design Centre had greater influence on the final product than would have been the case in other comparable companies. Often, as in the case of the early WM-2 model, the engineers would be required to follow the parameters set by the Design Centre, rather than the other way around (du Gay *et al.*, 1997: 63–4). There is no evidence that such a reversal in hierarchies has occurred in the development of other mobile devices. Nevertheless we suggest that examining the cultures of production, including the everyday experiences of designers, engineers and marketing strategists, is a fruitful way to conceptualise use, rather than a distraction from the issues of consumption. If part of the design of the Walkman was that it was to be a social (or at least shared) activity, what are the analogous assumptions which we are making about collaborative work practices?

Particularly where the designers use themselves as idealised early adopters of the technologies, as one of us did for the RIM pager study which we report here, it is productive to reflect upon the work practices which they (we) employ and cultural worlds which they (we) inhabit. An example of this is the extent to which the design of mobile technologies draws on models of traditional, stationary computing, literally taking the "desktop" out into the streets. Although it is clear that working with a PDA is not the same as working at the desktop, many mobile technology applications are scaled-down versions of traditional office applications.

11.2.1.4 Consumption

Cultures of consumption are in some regards the most obvious social process involved in creating the cultural history of an artefact. However, the process of consumption is not necessarily straightforward, and can rarely be predicted in advance. It is in use that the assumptions of designers may unravel or be diverted, just as was the case with the demise of the two-headphone socket feature on the Walkman. Sometimes the cultures and practices of consumption can be counter-intuitive for a mobile device industry focused on technical features and technical fixes. For example, in an informal study amongst students in the south-east of the UK, it was found that on the commuting train to and from campus, groups of young users spent more time talking *about* their phones (size, colour, features) or using non-networked services (games) than they did actually talking on the phone. The considerable amount of non-network use is underscored by the number of fake mobiles that are purchased and used as status symbols in social situations. Lycett and Dunbar (2000) noted such behaviour among young men in bars in the UK.

11.2.1.5 Regulation

The most well known issue around regulation of the Walkman was the diffusion of sound from a supposedly private device into the immediately surrounding environment. London Underground posters advised "Keep your personal stereo *personal*" and warned, "If your stereo annoys other passengers you are contravening the bye-laws". The image in Figure 11.2 shows a sign in a restaurant window. Some public arenas confiscate cell phones when one enters and there has been much discussion of technologically creating "silent" areas where cell phone calls cannot be received. Such social and legal sanctions can be traced back to the disruption of the public and private boundaries, which both kinds of mobile devices, in their different ways, allow, even in situations where bye-laws are not being broken.

11.2.2 Using the Circuit of Culture as an Analytic Tool in Design

We suggest that it is useful for designers to use this model reflexively when thinking about the practices into which their technology may or may not become embedded. However, we would like to note explicitly that the circuit of culture is

Figure 11.2 Regulating the use of a mobile technology through social sanctions
(Reproduced by kind permission of Lester D Nelson © 2001)

not a prescriptive model, but a descriptive one. As such it is a useful analytic tool in the design/production process and *not* an engineering model that will deliver design answers. The circuit of culture approach helped us to avoid thinking that the consumption of the mobile device was the only significant social process when attempting to devise new design frameworks. Our analysis further underscores the fact that successful mobile technology design is not simply a matter of specifying and realising technical functionality, and then considering appropriate interface "affordances" (Norman, 1988).

For example, we propose that advertisements play a role in the development of people's notions of how devices "should" be used. This is perhaps more noticeable when new technology genres are entering the market. Such advertisements draw on, and feed into processes of identification by specifying social uses. Such representations are designed to trigger recognition and desire ("if I buy a Walkman I must be/will be cool"). We further suggest that such representations feed into processes of design/production in the form of stories of use by imaginary consumers. Such stories are used to justify design decisions. Of course this process is a subtle and complex one. Fieldwork analyses can feed more stories of actual consumption practices into production, strengthening the links between consumption and production, and balancing the tacit effects of representation. However, only by explicitly addressing the embedded assumptions in common representations can we be aware of points at which those assumptions are playing out, and may be at odds with what we have observed in practice.

Our model is summarised in Figure 11.3. In the next section we describe some of our analyses of common representations of mobile technologies and technologies for mobility. Following this we present some fieldwork observations on the use of mobile technologies. In Section 11.6 we present a framework we have been using for the development of scenarios in design. Of course, Figure 11.3 itself presents an idealised and partial model of the design process; like the circuit of culture, it is not aimed at prescribing design practice but as offering heuristic guideline.

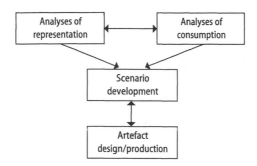

Figure 11.3 A model for analytic and empirical data gathering as part of
artefact design/production

11.3 Understanding Representations of Mobile Workers

Advertising featuring workplace use of mobile devices has focused on certain
types of business activities, constructed primarily in terms of availability. The
expectation that new technologies have or will find a "natural" home (and therefore market) within highly time-sensitive competitive exchanges has a long
history. The introduction of the landline telephone was initially promoted as a tool
for business men to keep up with trading exchanges (Marvin, 1988; Fisher, 1992).

In the case of mobile devices such advertising both draws upon, and creates
images fusing technology and availability (accompanied by the suggestion that
this is what is necessary to be a "good" businessperson.) The history of radio engineering reveals that in 1906, Lee de Forest, a radio engineer, transmitted a message
to an experimental car phone in New York. The president of De Forest radio was
quoted in the press as saying, "Hereafter, we hope it will be possible for business-
men, even while automobiling, to stay in constant touch" (Hijiya, 1992). The discourse of constant availability of people and of information through mobile
technology is not limited to the United States and Europe. Japan's largest cellular
phone company, which has 56% of the Japanese market and is probably the world's
largest cellular phone operator, is called DoCoMo, a nickname based on the
Japanese word for "anywhere".

In studying the mobile worker we have focused on recent representations
(January 2000–January 2001) of the mobile worker in the UK and the USA. We
acknowledge that representations of the mobile worker will vary across different
countries, cultures and times. Images of mobile workers appear on billboards, in
magazines and in other sales propaganda, appealing broadly to processes of identification. Notably, there is no clear set of attributes that provides a clear definition.
Social events and certain locations can provide the excuse to enrol potential converts. For example when London Underground (subway) workers went on strike in
the late 1990s, technology advertisers used the opportunity to highlight the potential advantages of moving the work base to your home, rather than having to struggle to your place of work.

Unsurprisingly, the mobile worker has been constructed as the audience *par excellence* for adverts in in-flight magazines like United Airlines' in-flight magazine, *Hemispheres*. One advertisement that appeared in the May 2000 issue advertises Sanyo's SCP4000 mobile phone with Sprint PCS as the service provider. A neatly manicured hand holds a cell phone. A 1.5 inch deep screen with six lines of text is shown offering a menu of options including "call history", "messages" and "phone book". The text alongside the image boldly declares: "*Your Office. Small enough to fit in the palm of your hand*". Smaller text beneath proclaims that this device is ahead of the competition because at only 4.7 ounces (agreeably portable for an office), it has an "easy view", "blue backlit display", and offers 180 minutes of talk time and 120 minutes of standby time with a standard battery.

Another advertisement in the same issue reads "*My hours, my ideas, my answers*" and shows a young man in shirt and tie brandishing a RIM two-way, interactive pager at the reader. The pager screen can be clearly read and displays the following email message: the subject header reads "Be in late" and the message goes on to say, "Thx for sending the info. I'll swing by their office to close the deal. See you at 10.30." The accompanying text for the advertisement states that you can "*exchange information when you want it, where you want it*".

In both of these advertisements ideas about connectivity are foregrounded, casting success as grounded in access to others ("talk time" and "messages") and to information ("when you want it, where you want it"). The January 2001 issue of the same magazine advertises the GoldMine Everywhere Server contrasting a PDA ("E-mobile customer management") with a desktop computer ("Immobile customer management"). The small print states that the GoldMine Everywhere Server is going to let "people tap into their information and work tools wherever they go. Communicate real-time within the sales group, and with customers. Co-ordinate efforts with online scheduling, lead management and team selling." Successful teamwork is cast simply as access to information.

One final example is again from the January 2001 *Hemisphere* magazine, and is for Sony's CLIÉ A photograph covering two pages shows an eager young man gesticulating and speaking intently across a restaurant table to an older man who sits to the right of the image, nonchalantly sipping from a tea/coffee cup. Screen shots hover beside their heads. Beside that of the younger man the screen reads "Career management, Effective E-commerce, Leadership Workshop, Business Feng Shui and Documents to Go". Beside the head of the older man the screen reads "Galaxy Invaders: Siege 2000". Here ambition meets (leisured) success. The text at the far right of the image states "He's young, ambitious. A real go-getter. Annoying. Just like you used to be. And he's got his strategy mapped out on his new CLIÉ handheld ...".

It would be unwise to generalise too far from this set of four advertisements, or indeed from the larger set of advertisements we have considered (12 advertisements in total). However, we have been able to observe certain quite specific idealisations of how the mobile device might be used. While these idealisations are clearly based in exaggerated fantasies that are aimed at selling specific technologies, following du Gay *et al.* we suggest they also contribute to a common discourse of mobility. This discourse, by necessity, promotes certain activities and values, while rendering many activities invisible. Representations show the "proper" use of mobile offices and a system of evaluating them (fitting in your hand), and help to set expectations of use. The activities associated with the devices are presented as unambiguously positive. In this way advertising becomes part of the "domestication" of the product in the lives of the business users, just as the home PC had to undergo a process of domestication (Silverstone and Hirsch, 1992).

In addition to explicitly stated goals – unlimited *access to others* and unlimited *access to information* – we have identified three key, interrelated features as being distinctive in the domestication of the mobile device as represented in these advertisements.

First, this kind of worker is supposed to possess a high level of control over their activities. This control is based upon interrelated capacities to act; these devices *allow* the independent, go-getting worker to be unshackled and therefore more effective, to be mobile. Such technology use is more than mere skilled practice of device use; control comes from the user in a specific alliance with the device. This kind of hybrid agency co-constructed between the human and the non-human has been explored elsewhere in science and technology studies, most clearly in Latour's study of Aramis, a Parisian rapid transit system (Latour, 1996). In the case of representations of the mobile worker it is not just that the device will allow one to *replicate* activities in the office in other spaces, but that one can *surpass* such activities with, and through, the device: "And he's got his strategy mapped out on his new CLIÉ handheld ...". In so doing, such images obscure other technologies and other formats in which our information is created and kept (e.g. paper).

The second feature of the mobile worker that is implied in these representations is that they are highly individualist at the point of use of the technology. This fits in with the competitive ethic of business. Although they may be collaborating at a distance on a document or a meeting, they do so through individual use of personally operated, private mobile devices (i.e. not accessing through cybercafé). In these readings, the technological devices themselves are not part of a collaborative network of devices and collaborations. The user effectively works alone, using the device that asserts and assures their independence. Notably, the devices themselves are not given equal status in terms of agency. Without a human user they cannot function.[2]

Third, these workers are embedded in a specific kind of temporality. The rhetoric of availability can be subdivided into several different subjective states: sense of urgency, vigilance about connectedness (activity of checking signal on mobile, checking email), awareness of time in terms of when last logged on (or similar) and the perspective that time is the ultimate limiting factor. In other words this is the promise that will get everything done – the only limit is time and not technological features. This kind of temporality is embedded in a particular version of the future in which there is an imperative to connect. In a study carried out in Manchester in the UK, Green and Harvey (1999) demonstrated that the "imperative to connect" circumscribed the actions at the level of individuals who wanted to connect, as well as promoting the agenda of local government sponsored projects (Electronic Village Halls for retraining of women and ethnic minorities). With mobile workers the imperative to connect is manifested in terms of imperative for mobile vigilance.

In the next section we describe some insights from field studies on the collaboration and mobility of people, devices and bits in the context of the themes identified above: access to others, access to information, agency, activity (both individual and co-operative), and temporality. We also address the way in which those themes are embodied in technologies and practice.

In moving to descriptions of the practices of mobile device consumption, we would like to make explicit that we do not believe designers draw on advertisements directly or consciously. Rather, as the circuit of culture makes clear, forms of representation like advertisements contribute to a general discourse of mobility. This generalised discourse *plays into* common narratives of use. Other sources of consumption that are not purposely manufactured are one's personal experiences and the experiences of others one knows or can observe. Narratives of use are also experienced through magazine and newspaper reports and through movie and television viewing. Our main point is that narratives of consumption are too often created from partial and fragmentary knowledge drawn unsystematically from multiple sources, and when applied in the production process in similarly fragmentary ways, can produce unwarranted assumptions about desirable device functionality and device design. We believe it is fruitful to turn to more systematic and focused analyses of consumption to create a discourse of mobility and mobile device use that is more directly relevant to people's experiences. We further believe that gaining an understanding of common rhetorics of use as embodied in cultural artefacts like advertisements can provide a useful analytic tool when devising field work investigations, allowing one to look for evidence of why someone purchased something, what they were expecting, what unexpected advantages they have observed, what disappointments they have experienced.

The above drove our analyses of fieldwork observations. Issues we have raised to addresses:

- access to information is unproblematic
- access to information is always desirable
- access to others is unproblematic
- access to others is always desirable
- access to us by others is always desirable
- devices will reduce the time taken to achieve our goals
- devices will reduce our time connect with others
- devices will increase our productivity.

Questions we have posed in design have included: What devices do we design if these assumptions are present? Are these assumptions borne out? Where they are not borne out, do these provide opportunities for design? How do people get around the issues? What other designs are possible?

11.4 Understanding the Practice of Mobile Work

With the increase of mobile technologies on the market, researchers have increasingly attempted to understand the experience of mobile work to determine precisely what aspects of mobile work are best and least well supported by current technologies. When characterising mobile work and mobile workers, certain categories of mobile worker are often invoked. These categories offer a snapshot of "typical" behaviours. Such categories are frequently used in marketing; within design processes they have served to underpin the development of detailed scenarios from which design ideas can be generated and within which they can be tested (Eldridge *et al.*, 2000; Lamming *et al.*, 2000). Such categories offer a sense of generality but stay in some sense close to activities that are supposedly shared among a group. Example categories include "road warriors", "globetrotters", "corporate wanderers" and "corridor cruisers". Road warriors are highly mobile, and are frequent fliers who spend a large amount of time on the move. They are seldom in a central location or office and are often in different countries and time zones. Globetrotters are also frequent fliers but typically are distinguished from road warriors by having a central office that is seen as their central base. Corporate wanderers are seen as spending a lot of time visiting others in their own companies, so these can be distinguished from the previous two in that they may well have access to familiar technologies and intranets. "Corridor cruisers" are local travellers who spend nearly all of their time within their own office building, but who are often not at their desks (and therefore not deemed to be as available). Bellotti and Bly (1996) offer a useful characterisation of this category of worker.

Another way of describing mobile work has been in terms of the consequences of mobility for collaboration rather than the degree of nature of the motion itself. Churchill and Bly (1999c) for example describe different kinds of mobility in terms of "presence", either in person or online, for collaborators. Framing the experience of the mobile worker through ideas of presence/absence of others highlights the importance of connectivity, both in social and in technical terms.

In our work we have found it is useful to disaggregate the concept of the mobile worker, moving away from categories of mobility-type or the extent of connectivity. Certainly, both of these are crucial components of the experience of mobile work, but an exploration of the local practices suggests that there is a whole array of experiences that do not easily come under these categories (see also Laurier, Chapter 4; Sherry and Salvador, Chapter 8; O'Hara *et al.*, Chapter 12).

Taking a closer look at all the tasks which may be involved in mobile work involves, amongst other things, detailing the actual practices of travelling, habits around acquiring and analysing particular types of information, considering the specific features of the spaces in which people are mobile, and the nature of people's communication ties. In expanding our definition of mobile work beyond depictions of people (typically men) closing deals and planning strategy in the advertisements, we have addressed both home and office roles, "desktop" technology use, in addition to analyses of mobile device use.

As noted above, our investigations drew on the concepts we drew from our analyses of advertisements; from the concepts we derived questions and looked for moments of "fracture" between advertised use and actual use. We looked in particular for instances of the themes identified: person and device independence; people's experience of time (collaborative rhythms); and the role of devices in making them feel more in control of their communications (a satisfied "imperative to connect") and their schedules.

11.4.1 Using Smart Pagers

Our study of pager use amongst two colleagues collaborating on software and hardware design shows that connectivity is a problem of asymmetry between devices (and devices/software versions) and reciprocity between people, rather than solely a case of network access. We also note a crucial distinction in people's concept of "email", which was constructed as an informal information repository, complete with attachments and embedded URLs, and the advertised concept of emails as informationally complete, self-contained messages.

The smart pagers were used over a 40-day period between December 1999 and January 2000. The pagers in question were the RIM Interactive 950 and the RIM Interactive 850. Logs were kept during the 40-day period of conversations including who with, when, content and length of thread (Marshall *et al.*, 2000). The main uses of the pagers can be summarised as follows:

- keeping in touch (sending short messages to establish and maintain communication)
- co-ordination (suggesting where and when to meet or talk)
- asking and answering short informational questions (e.g. asking about the whereabouts of items, asking for other's contact information, asking about project status)
- as a stripped-down surrogate for regular email (longer messages with multiple topics)
- device switching (proposing change in communication form, for example, pager to phone).

There were also some other less frequent uses. On one occasion no business card, pen or paper were available to enable conversants to exchange contact information, so the pager was used to send a message saying "This is where you can contact me". On another occasion the RIM 950 pager was used to send a message to the main work account so the email addresses could be easily integrated with the existing address book (this pager could not be synchronised with the main mail tool's contact list). Finally, emails were sometimes sent to oneself, to arrive at the main work email account as a reminder for the following day, much as answerphones are often used to leave self reminder messages. Although the pagers have web browsing capabilities, these functions were seldom used due to the small screen size, high connection latencies and the fact that most web pages are not designed for viewing on mobile devices.

In so far as many of these uses relied on the pager being connected to a wireless network, one of the central foci was an exploration of the experience of connectivity, and the concomitant sense of (advertised) "control" over one's work activities.

Within the pager study an awareness of basic network reliability was unavoidable when both users, as part of the study, moved from the 950 Interactive pagers to the 850 version of the same pager. This changeover occurred about 21 days into the use study. The 850 has increased functionality (it will synchronise with Microsoft Outlook), but relies on a different service provider and network from the

950. Unfortunately, the network service for the 850 was significantly less reliable than that for the 950; messages were not delivered in a timely fashion, and often required an explicit poll ("check messages") to see if messages were being held on the server. The 950, by contrast, was extremely reliable in delivering messages to the pager without any end-user action. The main reason for these problems appeared to be the way in which the network service providers had configured the mail servers.

The loss of reliability resulted in much frustration to message senders due to lack of information about if/when messages were received. The degree of availability and responsiveness during the use of the 950 pager had set an expectation for the rhythm of collaborative interactions on work projects. The subsequent loss of reliability as the new device was adopted disrupted these work rhythms significantly, and in the end resulted in the abandonment of the device altogether.

Such observations clearly undermine the notion that devices enable one *individual* control of one's communications and therefore one's work activities. Rather, the degree of dependence on the service provider's infrastructure became acutely evident; device utility was dependent on the configuration of the "support" infrastructure technologies over which there was no end-user control, and not simply within-device features. Further, the collaborative negotiation of ongoing work communications was very apparent; once the pager was no longer perceived as a predictable or reliable means of making contact, use declined fairly rapidly and a critical mass of contacts could not be sustained. Returning to the RIM 950 was not an option as collaborators had already migrated to other means of contact; these were largely cell phones and office-based email. In addition, far from connectivity simply being about access to the PCS network, the decrease in the use of the pagers was a social issue. Ongoing use is intimately related to being able to establish reliable communication rhythms with others.

A related issue was created by a lack of function symmetry: a phone is not a good communication device if you are the only person who has one.[3] There were clear instances of pager-to-cell-phone text messaging. Cell phone users were unable to send text messages back. As a result numerous one-way messages to colleagues' phones were sent to co-ordinate physical locations for face-to-face meetings. In addition there were asymmetries in colleagues' expectations about email/message response time, ability to read and share attachments and ability to read long messages, which often got truncated. This was due to the clear difference between email accessed on the pagers and regular use of email from a desktop machine: email at the desktop offers hyperlinked information (through embedded URLs) and is used as an informal repository of files in the form of attachments. These practices were not supported by the pager, somewhat belying the advertised claim "exchange information when you want it, where you want it". Although URLs could clearly be sent in email messages, accessing these via the pager was problematic due to the issues raised above. Thus, information which felt like it should be close at hand was in fact extremely hard to get at. Senders believed the email and the attachment had been received, recipients were frustrated by being unable to open files and follow URLs. People were clearly more frustrated about not being able to access information that they felt they *should* be able to get at, that is information which, had they been at the desktop or using email on a laptop, they would have easily been able to read. Thus information was differentially valued, and not having access to it led to highly different tolerance for lack of access.

This is reminiscent of a distinction proposed by Millar *et al.* (2000) between information that is close and that which is perceived as being "distant". They derived this distinction from the observation that people's ideas about the data they were retrieving revealed a notion of "close" information and "distant" information, even though in a network sense these terms have little meaning. These terms can be thought of in the following way. "Close" information is that which is in some sense local, and is perceived as being readily available. Examples include what was on at the local cinema, local government information, even information about familiar consumer products. "Distant" information by contrast is in some sense less usual or familiar. Such information may be about foreign universities, far off experiences/people, and remote holiday locations. This conception led to the interesting observation that, to the user, closer information should be "fast", whilst distant information might be tolerated as "slower" or "poorer quality" because it comes from afar.

Another practice was the use of multiple devices (laptop, cell phone, pager, notebooks) to cover the potential problems of technical glitches and because necessary information was often spread across multiple devices. There was a form of "device collaging": if I have my laptop, my pager, my paper notebook and my cell phone, I have all my contact and all my calendar and work related information. I also have a better chance of being connected. We suggest that such device collaging is a common practice for many of those who work with technology, whether or not that technology is mobile. Suchman and her colleagues have described a similar phenomenon using the concept of "archaeological layering" of technologies (Suchman *et al.*, 1999). Their image of available technologies in the operations room of an airport resonates with the experience of using the RIM pagers within the device collage:

> Rather than being homogeneously and seamlessly integrated, these artefacts comprised a heterogeneous collection of information and communication technologies, including telephones, radios, video monitors, networked workstations, whiteboards, clocks and a wide array of documents. The integration of these artefacts, correspondingly, seemed more a matter of string and bailing wire than of design. (*ibid.*: 397)

In sum, the pagers were only as good as the entities with which they could be synchronised – in terms of both technical and social network synchrony. Our observations underscore the importance of predictability and reliability in the building of collaborative rhythms of work: a reliable communications infrastructure; good connectivity; critical mass; and symmetric device capabilities enabling reciprocity. Finally, we note that not all information is valued equally. People's expectations about access to information and their frustration when they could not get access could not always be evaluated in terms of people's current work goals. Rather, people often got frustrated about not being able to get to things that they were used to accessing without problems. "If it's supposed to be email, then I should be able to open my attachments!" Therefore rather than thinking of all information as equal we propose a continuum along which information can be placed. This continuum is one of closeness or distance, and is evaluated in terms of people's current goals and desires. When I am travelling, a book on my office shelf is only evaluated by me as distant when I want it, but can't get to it. A file on the server at work is close when I can connect and I want to read it but distant when I can't get my modem to connect in from my hotel room. A file on my laptop may be physically close to me, but is really distant if my battery dies. And a paper document may be close to my face but the content is distant if it is all written in a language I don't understand.

Interestingly our notes above on closeness and distance of information are in collaborative working relationships intertwined with the nature of our working relationships. We also noted a distinction that is common in CSCW but is not raised in our readings of the rhetorics of devices. That is, that relationships are also not on or off. We have different working relationships that can be characterised as tightly and loosely coupled. It was clear in our study here that this interacts with closeness and distance. If I want access to some information and I am working in a tightly coupled way with someone who has access to that information, I can call them and ask for it. We see this scenario in our analyses of cell phone use described below. However, if I am not in a tightly coupled relationship with someone, I cannot get at distant information with their help. This relationship continuum subsumes many of our themes of agency, independence and temporality. However, these themes are crucial for understanding the way in which the tight and loose coupling is negotiated. In the next section we continue our analysis of these themes in the context of cell phone use in the UK.

11.4.2 Use of Cell Phones in the UK

A study of cell phone use in the UK again showed the diversity of ways in which mobile devices are used to create and sustain work and leisure relationships (Churchill *et al.*, 2001).

Eleven people were interviewed and observed over a period of between one and seven hours. Interviewees included three teenagers, a tax consultant, a district nurse, a freelance documentary photographer, a research scientist (whose work was based at three different research institutes), a food store manager and a secondary school head teacher. Length of time of owning a mobile phone varied from one week to several years. Although the central focus of the interviews was cell phone use, given our interest in device collaging, other communication technologies such as laptops, landline phones, letters and PDAs were also discussed. Interviewees were also asked to reflect on their closest collaborations and most frequent communications in face-to-face situations as well as using communication technologies.

These interviews revealed a great variation in people's degree and types of mobility, and in the ways in which they maintained (or controlled) their central working and leisure relationships while remote. A central observation was that the cell phone was simply one of many means of keeping in contact with others. In all cases, interviewees had at least two phone numbers associated with them (landline and cell phone), had voice mail and answerphones, some had email and most had both work and home addresses. Again this is clearly a form of communication device/media collaging. A cell phone in most instances was used to supplement face-to-face interactions.

The extent to which ownership of a cell phone was deemed essential in this lattice of communication media varied depending on:

- the extent to which people felt others were dependent on urgently contacting them
- the extent to which people's locations and mobility were predictable
- the amount of time they spent in locations where the cell phone was the only means of contact.

People's perceptions about these factors were not easy to establish from the outside. Certainly establishing the degree to which people considered their cell phones crucial did not correlate entirely with any established stereotypes about occupation. The food store manager, for example, felt her cell phone was essential, despite being highly routinised in her schedule and having almost constant access to a landline phone (as there was one in the shop and one at her home). She said she felt a strong need to be in touch with her friends during the day, and often used text messaging from her phone as well using the phone to talk. Communications she deemed urgent and essential were usually about social arrangements and about occurrences during the day. She used her cell phone rather than the shop landline phone so as not to "tie up" the shop phone. This was clearly about perceived social "pressure"; use of the shop phone was more "visible" than use of her personal cell phone, and might lead her to be overheard and socially accountable to others.

Similarly, people's perceptions were often hard to read as control of the device was in some sense out of their hands – again belying the notion that mobile devices consolidate and strengthen people's autonomy. Indeed, outside "authorities" determined to some extent their use of the cell phones. For the teenagers, this urgency about contact was focused on their parents knowing where they were; their schedules were seemingly highly predictable but nevertheless the family convention was to maintain tight monitoring. Green notes in Chapter 3 in this volume on mobile devices and surveillance, that mobile devices have fuelled "the notion that individuals should be available and accountable to others, visibly and transparently at any time and place", and notes issues that occur when competing authorities (school authorities and parents) are brought into direct conflict – "turn the phone off while at school", "keep the phone on so I can get hold of you". This accountability is also manifest at bill paying time; for all the teenagers, their parents paid the bills. The bills and all numbers called are therefore available for careful scrutiny.

"Watching" of this kind clearly takes place for many workers who are provided with cell phones by their companies, turning a communications device into a surveillance mechanism – affecting their use of the device through the setting of externally determined boundaries of acceptable use, and requiring judgements about permitted and non-permitted communications. It was evident that being available to one's boss via the cell phone could sometimes be at odds with the requirements of the social situation in which one was physically located.

In this vein, the district nurse who was highly mobile, often in her car and working in a domain where one might expect her to be "on call", did not feel personally that she should be under pressure to be available. For her the cell phone was both a site of convenience (pleasure) and pain; it was good that she could be contacted by her family, but there was a certain tyranny to feeling the phone should be always on. This led to a certain pleasure when connectivity broke down for technical reasons that could be construed as "not my fault", e.g. loss of network connectivity or battery problems. She could enjoy the "seam" that was created between herself and her direct managers. Specifically, her current cases meant she was unlikely to be called on "life and death" emergencies. However, her estimation of the purpose for the cell phone – "real" emergencies – and her immediate superior's notion of the purpose for the cell phone were clearly at odds. Too often cell phone calls were to give her additional administrative tasks.

For others such breakdowns were an entirely negative occurrence. While the district nurse's location accountability was to her work authorities, for the freelance photographer this pressure was in terms of speed of responsiveness to current and potential clients. If he were not available, this could mean someone else (who was available) would get the job.

The extent to which such accountability poses problems depends on the social support infrastructure. For the tax consultant, who travels regularly, a source of frustration was that his cell phone worked when he travelled in Europe, but did not work when he was visiting sites in the USA. This required careful handling: often he would warn clients to expect less timely responses from him. This kind of work had increased significantly since he started being regularly available on his cell phone. Prior to that not responding to requests for a few days had not required explanation. However, the nature of his workplace support infrastructure (having a secretary) meant he was available in some sense (through his secretary as his proxy) and therefore did not feel the pressure to be available at all times himself. This option was not available to most of the interviewees so the cell phone took on a different significance.

Cost was also an interesting point of note. For the teenagers, sending text messages from their phones was far cheaper than calling. Text messaging was therefore the main mode of communication for these teenagers; phone calls were reserved for their parents. Further, text messaging was free (at the time) for those on the same network, but had a price attached if crossing networks. This led to a partitioning of friends into those who were "close" (free to contact) and those who were "distant" (cost too much to contact). This notion of closeness and distance corresponds with analyses of information posited by Millar *et al.* (2000) and will be raised again later in this chapter. The main point here was that all friends and all communications were not equal, but were differentially accessible (or not) as determined by the network costing structure – and not the technology's functionality.

These observations again highlight the fact that feeling in control of one's work and leisure relationships is more than accessing information or people whenever needed. Further, the social and technical infrastructure issues belie the notion of the individual in control of their own destiny or the device as the sole tool for all jobs. We can see the technical solution reified in so-called convergence devices. Here, the driving force appears to be to "provide all the functionality that is technically possible in one device and the entire user's needs will be met". Evidently, life is not so simple.

11.5 Broadening the Concept of Mobile Work: Reflections on Fieldwork

The observations above reveal a much more complicated notion of collaborative work and the role of mobile devices in support of that work than is often assumed. Returning to the themes identified above we can begin to address the capacity to act with mobile devices, the nature of collaboration on the move and people's construction of temporality.

First the dependence of mobile connections on stable infrastructures and the seamless move between mobile and "static" modes of working is evident. The mobile individual in control of their own time and working relationships is supported by a social, environmental and technical infrastructure. The need to be reliable is paramount, and the use of any particular technology is subsumed under that. Examples from our previous research on virtual environments (Churchill and Bly, 1999a, b, c) and cybercafés (Wakeford, 1999) demonstrate the importance of a stable infrastructure in conjunction with which people can be mobile.

Whereas in advertising discourse the capacity to act when using a mobile device was linked to key technical features of that device, we found that there were social conventions of communication which prevented straightforward collaborative acts, such as returning a pager message or tailoring a phone call to the suitability of location in which it was taking place.

Furthermore, connectivity to the network became the key sticking point only when the network failed. When people did "lose signal" it often revealed how little they knew about (and far less controlled) the infrastructure. As a recent study of the 1998 California Galaxy IV Satellite blackout demonstrated, much of the infrastructure only becomes apparent to people when it fails (Hong *et al.*, 1999). There was some evidence in our study that people anticipate this invisibility through device collaging rather than attributing to one device the capacity to overcome the fragilities of infrastructure.

The hidden work of conducting activities while being mobile was also quickly apparent in fieldwork situations. Practices revolved around waiting for events to happen, such as flights, information on business deals, or even mundane waiting for a call back on the telephone. Waiting for other people to do things was a common experience which demonstrated an interconnected network of working relationships far from the image of the sole mobile worker of the advertisements.

The fact was however well your technical device functioned, you were still crucially dependent on your collaborators and also the interconnectedness of devices.

We suggest that the appearance of infrastructure and the kinds of hidden work necessary to sustain collaborative work while on the move might lead us to look at what part of the device collaging must be stable for the mobile devices to function. This leads us to go beyond the interviews with those who use mobile devices to examine the kinds of stability that underpins collaborative use of those devices.

Finally, from our characterisation of mobile collaboration we have put the concept of mobility under closer scrutiny. Although mobility is usually associated with movement between geographically situated spaces, and hence is linked to travel, it must also be thought of in terms of time and not simply as expedient goal achievement. Mobile workers do not just need to connect while travelling (in airports, on trains, in cars, at remote locations). Their fundamental experience of mobility is embedded in an experience of temporality which includes mutually negotiated rhythms of contact, availability and accessibility. Cultural geographers have long indicated that space/time cannot be as easily separated as is suggested by some of the contemporary use of mobility as a way to describe the central experience of these workers (e.g. Massey, 1992; McDowell, 1999).

Two central themes emerge: access to information and access to others. These themes were clearly evident in discourses of mobility that we noted above in our analysis of representation. However, where we observed access to others and to information as being unproblematic in predominant discourses of mobile device use in advertising, we found that this is indeed problematic. Problems arise from technical problems, from personal preferences and from an inability to connect with others due to scheduling difficulties – sometimes caused by the devices themselves. Secondly, temporal and scheduling issues were of great interest.

11.6 Putting it all Together: Design Frameworks for Collaboration on the Move

So how are we to summarise our findings – both the analyses of representation and the analyses of consumption? In this section we describe the dimensions we have developed to encapsulate our findings and that have helped us better understand mobile work practices. These dimensions have proven invaluable in our own technology design discussions, and have reportedly been used by others in their consideration of mobile work (John Sherry, INTEL, personal communication). The dimensions are shown in Figure 11.4.

If the promise of technologies is to make all distant information close, one of our observations was that people differentially value information, and the desire to have something close cannot always be predicted. Thus, when people were on the road certain pieces of information would become desirable, information that was not expected to be needed. In terms of mobility, again we observed a certain fluidity that was not embodied in many of the devices.

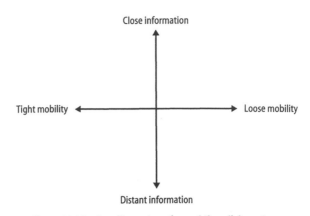

Figure 11.4 Design dimensions for mobile collaborations

As noted, mobility must be linked with ideas of temporality and synchrony of communication. Therefore, we suggest that this can be undertaken by the development of an analytic device of a continuum. This continuum runs between "tight mobility" and "loose mobility". Tight mobility is the experience of needing real time synchrony while on the move. It is maintained through ongoing negotiations in established relationships where location information is easily shared or predicted, and is therefore highly collaborative. Loose mobility is the requirement of accessing documents or information on the move, but asynchronously, and does not require input at such a detailed level. It is therefore highly co-operative, but not collaborative at a detailed level. An example of tight mobility would be using a mobile phone to conduct a conference call. An example of loose mobility would be accessing and editing collaborative documents in another country.

It should be pointed out that the real time "chat" implied by tight mobility does not need to be voice only. In text-based virtual environments synchronous, text-based communication is a key feature in determining the environment as a successful collaboration technology (Churchill and Bly, 1999a, b, c). Short text messaging on mobile phones can, when connections are reliable, simulate such real-time chat. Similarly, the use of voice mail over mobile phones or accessing voice mail from an office phone remotely would be an example of non-text based loose mobility. Interestingly, online, shared calendaring systems assume that day-by-day, hour-by-hour, and publicly visible, scheduling can *create* or encourage tight mobility. Informal evidence suggests however, that scheduling meetings with online calendars does not *create* tight mobility or tight collaboration where it does not already exist. Subversion of the technology includes blocking out days at a time for fictitious meetings.

A second dimension is necessary to capture the kind of cultures of information that are embedded in the experience of knowledge work. We propose a distinction of close versus distant information based on the analyses presented by Millar *et al.* (2000) discussed above. For the purposes of our design framework, however, we consider close information to be defined as information which can be gathered through the mobile technologies and other sources. Distant information might be defined as information which is constructed as "difficult" to access by other means than the mobile technology. For example, close information might include train timetables, or information related to collaborative work that can easily be accessed remotely (e.g. via the web at an internet café, or via a phone recorded message). Distant information may be stored on repositories that are not easily accessible (behind firewalls for example) or which are not integrated digital materials. The discussion above shows the relevance of piles of paper, Post it Notes and so on for reminding and for structuring people's time in and out of the workplace. Design for mobile workers and the development of mobile technologies must acknowledge these two dimensions of knowledge work.

We have used these dimensions in several design brainstorming sessions around the design of innovative mobile technologies. To date, these dimensions have led us to consider potential users' activities as a whole, rather than focusing simply on the situation where mobility is most evident – that is when they are actually moving. Most journeys involve moments of mobility, interspersed with moments of being stationary. Those stationary moments may be in locations where "wired" rather than wireless mobile devices are available. Thus, we can see that devices to be truly useful need to span the four quadrants as people move through these, and that support for fluid activity/situation changes is of great importance. As we develop scenarios using the dimensions we go from generalities ("tight mobility") to specifics of people's lives and to the ways in which issues like agency, individualism and temporal rhythms play out.

So for example, take the district nurse introduced above. Her day begins "low mobility"; she does not need to be in tight collaboration with anyone. She gets to the medical practice, then she moves to the top right quadrant. She is also moving left as she sets the agenda for the day. She is now face to face, tightly mobile (still not planning on staying at the office long) and close to information. She leaves on her rounds, and is still tight mobility as her supervisor knows where she is and her mobile phone is on. The information she needs is close. She gets to a client's residence and then realises she has forgotten something. This information now becomes relevant but is distant. If she could log on she could get it, if it is a digital file. Otherwise, it remains distant. We can now begin to speculate what kinds of things for different jobs may need to be digital. Even if she could log on, she may not be able to view the document from a small device. The information remains distant because of its format or the size of the file.

An example would be an overseas trip with a laptop and a cell phone as the main mobile devices. Prior to leaving on the trip one can imagine a person collaborating on the phone from home to load all the relevant information onto the laptop using a DSL connection. In this instance, the digital information is close and the collaboration tight. As the journey begins, information both digital and paper-based becomes distant as no connection is available. Mobility becomes looser if one is for example on a flight where wireless devices are not allowed. As one arrives at the destination, one may be able to re-establish tight mobility, if the wireless device works abroad. Information will still be distant if one cannot get access to a high bandwidth line.

From analyses such as these, we have identified several issues that should be addressed in mobile device design. Therefore the mobile devices we are designing need to:

- be reliable in different and changing locations, or provide early warning of lack of reliability, that is degrade gracefully
- allow for careful, more nuanced management of interactions given situational and personal constraints
- provide feedback as to where connection problems derive from, that is offer meta information about connection and rendering failures
- be more easily interoperable with other devices
- allow for management of different kinds of information, both close and distant – as determined by the user
- be able to render more types of documents
- enable fluid transitions between synchronous and asynchronous collaborative work over content through mobile devices.

All these points emphasise the interrelationships between work activities in the four quadrants of the design framework. Recent work has addressed point 2 with devices that can offer situation tailored responses by call recipients on cell phones (Nelson *et al.*, 2000), and points 5, 6 and 7 are addressed by two recently designed systems. These systems allow the sharing and multimedia annotation of digital content through mobile devices (Churchill *et al.*, 2001).

11.7 Conclusions and Summary

As noted above, the Sony Walkman was designed with two headphone sockets for joint listening. Over time the second socket disappeared as it became apparent that people were not using it. As we have seen with mobile devices, if you assume all devices are used alone, you don't design for interoperability. Our motivation in this chapter has been to propose that a wider range of stories of use find their way into technological designs. We also argue that cultural analyses can be usefully employed as a complement to detailed fieldwork analyses that look at local practice. Such cultural analyses can be used not only to place activity in a broader cultural framework, but also to drive specific questions within the fieldwork. We believe it can also help us to position novel technologies into a space of "market" readiness.

In this chapter we have provided an analysis of mobile work both from the perspective of representation within our culture and also in terms of people's everyday practices. Our analysis shows that notions of mobility and of mobile work often diverge significantly from people's actual experiences. Although not surprising in itself, this fact is significant for designers and once again argues for a better understanding of practice in the design process, especially in today's climate where technology innovation is moving fast and is driving day-by-day revolutions in expectations and practices of work. We argue that an understanding of practice needs to consider people's actual practices with and without technologies, but we argue for opening up the scope of analysis to consideration of the ways in which technologies are represented and use is narrativised within different cultures.

We further proposed that common notions of the mobile worker can be disaggregated into separate local activities, such that a deeper understanding of people's interactions with and through these devices can be understood. We offered a contrast between the representations that promise agency and control and of the individuality of the device and the worker and the experiences of people with their devices "in the wild". We illustrated how our mobile devices are dependent on network infrastructure and on interoperability with other devices. Our actions and interactions, in turn, are dependent on a rich and complicated social infrastructure. Finally, our fieldwork shows that we also need to disaggregate the notion of "information" in this highly promoted mobile information sharing dream. Notions of distance/closeness demonstrate clearly that we need to better understand how people differentially value information. Such understanding can only come from seeing the role of that information in people's working practices.

From these analyses we have derived a framework for the design of innovative mobile technologies; this framework emphasises not mobility per se, but rather the relationship between the nature of the collaborative relationship, and therefore the need for timely access to others, and people's requirements for access to information. This framework has been used to drive the development of scenarios for design and has been used as an intermediate representation in multidisciplinary design discussions.

As a final comment, we invite designers to engage with us on issues related to the ways in which designs are constructed and realised. Our aim has been to demonstrate the ways in which we can encourage creative engagement with technology design that acknowledges processes of representation and consumption.

Acknowledgements

We would like to thank all those who have taken part in our various studies, and our readers for their comments. We would also like to thank Les Nelson for the photograph of the restaurant window sign.

Notes

1. This term is taken from the work of Pierre Bourdieu (1984).
2. Although in some histories of technological innovation (e.g. Latour, 1996) actants are seen as hybrid forms.
3. It is of note that, even within device "families", there are often asymmetries caused by different models and the capabilities of different software versions.

References

Bellotti V and Bly S (1996) Walking away from the desktop computer: distributed collaboration and mobility in a product design team. In *Proceedings of the ACM 1996 Conference on Computer supported cooperative work*, 1996, pp. 209–218.

Bourdieu P (1984) *Distinction.* (trans. R. Nice). London: Routledge.

Carroll JM (1995) *Scenario-based design: envisioning work and technology in system development.* Chichester: John Wiley.

Churchill EF (2001) Getting About a Bit: Mobile Technologies and Mobile Conversations in the UK. *FXPAL internal technical report,* FXPAL-TR-01-009, 2001.

Churchill EF and Bly S (1999a) Virtual Environments at Work: ongoing use of MUDs in the Workplace. *Proceedings of WACC'99,* San Francisco, CA: ACM Press, pp. 99–108.

Churchill EF and Bly S (1999b) It's all in the words: supporting work activities with lightweight tools. In *Proceedings of Group'99,* Phoenix, AZ: ACM Press, 1999.

Churchill EF and Bly S (1999c) *Ubiquitous access to others: maintaining co-presence through MUD locales.* Workshop position paper, Workshop on Ubiquitous Virtual Environments, ECSCW '99, September 13, Copenhagen, 1999. Also available as FX Palo Alto Lab. Technical Report FXPAL-TR-99-024, 1999.

Churchill EF, Trevor J and Cubranic D (2001) MediaNote: collaborating over content while on the move. *Internal technical report,* FXPAL-TR-01-010.

Du Gay P, Hall S, Janes L, Mackay H and Negus K (1997) *Doing cultural studies: the story of the Sony Walkman.* Buckingham: Open University Press.

Ehn P (1988) *Work-oriented design in compute artifacts.* Swedish Center of Working Life. Stockholm: Lawrence Erlbaum.

Eldridge M, Lamming M, Flynn M, Jones C and Pendlebury D (2000) Studies of Mobile Document Work and their Contributions to the Satchel Project. In *Personal Technologies,* Vol. 4 (2 & 3), June 2000.

Erickson T (1995) Notes on design practice: stories and prototypes as catalysts for communication. In JM Carroll (ed.). *Scenario-based design. Envisioning work in technology and system development,* pp. 37–58. New York: John Wiley.

Fisher C (1992) *America calling: A social history of the telephone.* Berkeley, CA: University of California Press.

Green S and Harvey P (1999) *Scaling place and networks: an ethnography of ICT "innovation" in Manchester.* Presented to Internet and Ethnography Conference, University of Hull, December. Available at http://les.man.ac.uk/sa/Virtsoc/Scale.htm.

Hijiya J (1992) *Lee de Forest and the fatherhood of radio.* Bethlehem, PA: Lehigh University Press.

Hong T, Matei S, and Dutton WH (1999) *Missing the Pager: The impact of the Galaxy IV Satellite Blackout.* Paper presented at the annual meeting of the International Communication Association, San Francisco.

Kincaid CM, Dupont PB and Kaye AR (1985) Electronic calendars in the office: an assessment of user needs and current technology. *ACM Transactions on Information Systems,* 3 (1), pp. 89–102.

Lamming M, Eldridge M, Flynn M, Jones C and Pendlebury D (2000) Satchel: providing access to any document, any time, anywhere. *TOCHI* 7 (3), pp. 322–352.

Latour B 1996 *Aramis or the love of technology.* Cambridge, MA: Harvard University Press.

Luff P, Hindmarsh J and Heath C (2000) *Workplace studies. Recovering work practice and informing system design.* Cambridge: Cambridge University Press.

Lycett JE and Dunbar RIM (2000) Mobile phones as lekking devices among human males. *Human Nature,* 11 (1), pp. 93–104.

McDowell L (1999) *Gender, identity and place. Understanding feminist geographies.* Minneapolis, MN: University of Minnesota Press.

Marshall CC, Churchill EF and Trevor J (2000) Two-way pagers: our experience with dedicated mobile text chat communication. *FXPAL internal report,* FXPAL-TR-00-012.

Marvin C (1988) *When old technologies were new: thinking about communications in the late nineteenth century.* Oxford: Oxford University Press.

Massey D (1992) Politics and space/time. *New Left Review,* 196, pp. 65–84.

Millar C, Hunt M, Heath J, Macfadden M, Wakeford N and Haggett C (2000) Accessing local authority services using the internet-on-TV: results from a field trial in Suffolk and London. *Final Report to European Union GALA project.*

Nelson L, Bly S and Sokoler T (2001) Quiet Calls: Talking Silently on Mobile Phones, *Proceedings of the CHI 2001 Conference on Human Factors in Computing Systems,* Seattle, WA: ACM Press, 2001.

Norman D (1988) *The psychology of everyday things.* London: HarperCollins.

Palen L (1999) Social, individual and technological issues for groupware calendar systems. In *Proceedings of CHI 99 conference on Human Factors in Computing Systems*, 1999, pp. 17–24.

Palen L, Salzman M and Youngs E (2000) Going wireless: behaviour and practice of new mobile phone users. In *Proceedings of ACM Conference on Computer Supported Cooperative Work, CSCW 2000*, pp. 201–210.

Silverstone R and Hirsh E (1992) *Consuming technologies: media and information in domestic spaces.* London: Routledge.

Simonsen J and Kensing F (1997) Using ethnography in contextual design. *Communications of the ACM*, 40 (7), pp. 82–88.

Sivulka, J (1998) *Soap, sex and cigarettes. A cultural history of American advertising.* Belmont, CA: Wadsworth Publishing.

Suchman L (1987) *Plans and situated actions.* Cambridge University Press.

Suchman L, Blomberg J, Orr J and Trigg R (1999) Reconstructing technologies as social practice. In P Lyman and N Wakeford (eds). Analyzing Virtual Societies: New Directions in Methodology. *American Behavioural Scientist*, 43 (3) November/December 1999, pp. 392–408.

Wakeford NS (1999) Gender and the landscapes of computing at an internet café. In M Crang, P Crang and J Dey (eds.). *Virtual geographies: bodies, spaces, relations*, pp. 178–201. London: Routledge.

Chapter 12
Exploring the Relationship between Mobile Phone and Document Activity during Business Travel

Kenton O'Hara, The Appliance Studio, Mark Perry, Department of Information Systems and Computing, Brunel University, Bristol, England, Abigail Sellen, and Barry Brown, Department of Computer Science, University of Glasgow, Scotland

12.1 Introduction

The past few years have seen a startling boom in the uptake of mobile phone technology with sales far exceeding even the most optimistic predictions. This uptake is widespread across user groups, but in particular it has shown its value for the mobile worker. Many information technology (IT) companies want to leverage this fact and develop new devices and services which will take the mobile phone in new directions, or which will exploit the fact that so many people already carry them. So, for example, we are already seeing the emergence of new kinds of data services where mobile workers can remotely access, send and receive documents and information.

The potential for combining phone use with other kinds of information-related activities is beginning to raise important questions for IT and telecommunications companies. For example, how and why should wireless communications be incorporated into mobile devices (such as PDAs) to allow people to create, display or manipulate information while on the move? What new data services might be useful to access via mobile phones? What document handling facilities should be incorporated into mobile phones? In order to answer such questions, we believe a user-centred approach is valuable in order to identify unmet user needs, and as a way of inspiring innovative design ideas for both products and services.

12.2 Our Approach

Our specific approach is to look for new design directions by studying the potential users of new technologies and their work practices. In the study we report here, we chose to look at mobile professionals and their work practices during business travel. One reason for this is that mobile professionals are an obvious target for many of the new work-related mobile devices that are being developed. But in addition, we have carried out studies of many different kinds of office-based workers over the years (e.g. Puuronen and Savolainen, 1997). We wanted to focus this time on workers on the move to find out in what way their needs are similar or different from their office-based counterparts.

In particular, this chapter centres on specific aspects of the findings concerning their mobile communication and document activities (be they paper or electronic). In their reliance on talk and on text, mobile workers are no different from other kinds of workers we have studied over the years. However, as we will discuss, we found that the work practices of mobile professionals demonstrate some unique and interesting features of this relationship that office-based workers do not.

12.3 Previous Research

User-centred research of mobile work is only now really beginning to emerge as an important field in its own right. Despite the growing importance of mobile phone technology there has, to date, been relatively little research on their use, with a few exceptions (the contributions to this book notwithstanding). There have also been some important in-depth studies of mobile professionals in relation to their document activities (e.g. Eldridge *et al.*, 1999 and Luff and Heath, 1998). However, there has been very little work explicitly looking at the relationship between mobile phone use and the document activities of mobile professionals while on business travel. Whittaker *et al.* (1994) make some allusion to the role of documents as a conversational resource while on the move as an interesting theoretical issue to consider but they do not really develop it beyond the suggestion that it would be an interesting area to consider for future thought. Luff and Heath (1998) also present an interesting work practice example of remotely mobile collaborative work involving screen-based documents and wireless communication but again the concern of the paper is not explicitly with the relationship between mobile phone and document use.

12.4 The Study

On the basis of previous work such as segmentation studies of mobile professionals (such as the BIS study (Ablondi and Elliott, 1993)), we were well aware of the diversity in the nature of mobile work, the ways in which they travel, and their documents and technology use. A pool of participants was collected from which 17 mobile professionals were chosen to provide a representative sample across a range of different professions. They were as follows: regional manager for a market research company; corporate relations manager for a communications firm; regional operations manager for a telecommunications company; software sales manager; managing director of an Italian import company; account development manager for a major brewery; business and sales manager for a lab equipment supplier; PR consultant; medical research co-ordinator in a large hospital; international customer services manager for a telecommunications company;

international marketing director; civil servant (executive officer for procurement); sales and marketing manager for a software company; production manager for a television company; strategic account manager for the product support division of a computing company; business development manager of a research lab; project manager in an e-business application department.

The study used a combination of diary techniques, interviews, and analysis of the technologies and documents used during specific business trips. This allowed us to ground the study in real activities and to use trip diaries and existing artefacts to unearth the detailed context of their document and communication activities. Another important aspect to the approach was to gather data both before and after actual business trips so that we could gain a longer term understanding of what happened during travel: what kind of preparations were made prior to travel, and what actually transpired during the travel episode. This helped to give a deeper understanding of the context surrounding the mobile professionals' activities.

Each participant was interviewed twice. The first – the pre-travel interview – was conducted as close as possible prior to the departure date for the next upcoming business trip. The second – the post-travel interview – was conducted as soon as possible after returning.

12.4.1 Pre-travel Interview

This was divided into two parts. The first part was to gain some general background about the nature of the participant's work and home lives. This included: information about their position and responsibilities in the workplace; who they worked with and the nature and frequency of this collaboration; descriptions of typical days, both office and travel based; frequency and type of travel; information about family and social networks; technology infrastructure at home and office; and technologies used while mobile.

The second part of the interview focused more specifically on the upcoming business trip. This included: the purpose of the trip; trip duration; where they were going; who they were going to meet; what kinds of information they were expecting to gather; preparations they had to do, such as slide creation; what documents they would be taking with them and why; what technologies they would be taking with them.

Finally, participants were asked to bring with them to the second interview any materials (e.g. emails, paper and electronic documents) generated or collected while away.

12.4.2 Post-travel Interview

This was divided into three parts. The first part of the interview focused on gaining an overview of what had happened during the whole trip.

In the second part, the interviewers used the overview to decide on a typical day in the trip to unpack in more detail. A "diary" of this day was then constructed from the beginning of the day when the participant awoke right through to the end of the day when the subject retired to bed. For each "diary day" participants were asked to give detailed descriptions of their activities throughout the course of the day including any communication episodes, dealings with documents (paper or electronic), meetings and travel. Any information that was gathered or distributed (e.g. documents or photographs) was then photographed or photocopied. These artefacts were discussed in relation to the daily activities and the goals and motivations of the participants. Participants were asked what they planned to do with this captured information on their return from travelling.

Finally, participants were asked about any particular problems that they experienced in relation to their document-related and communication activities while travelling. Specifically they were asked whether there had been times when they *would have liked* to have been able to carry out some action but did not have the means.

12.5 Findings

In this section, we investigate the different aspects of communication activities across participants as well as their document related activities. We then aim to bring these together to look at the interaction between document activities and phone use, and present a taxonomy with illustrative examples of participants' activities when they were mobile.

12.5.1 Phone

All of the participants bar one took a mobile phone with them. Reliance on the phone for communications was indicated by participants receiving on average five phone calls a day. These incoming calls relied most heavily on the mobile phone with about 85% of calls being received this way. In terms of outgoing communications, participants reported making approximately 10 phone calls a day, although this varied widely, with some participants making up to 19 calls a day and others making only a single call. Of these, 55% of outgoing phone calls were made on a mobile phone.

One of the implications of this is that these business travellers were much more likely to be the initiators of communication than to be receivers. We found this particularly to be the case for more senior managers. Some of the senior managers saw outgoing calls as the major role for their mobile phone and made deliberate attempts to avoid incoming calls by, for example, switching off their phone or deliberately not giving out their mobile telephone numbers. The data thus confirm the dependence of these mobile workers on their mobile phones. As our television production manager described it:

> If you're in a studio then yes there's more of a routine but if I'm on location then I'm anywhere and then basically my tool is my mobile, that's my number one thing.

As the data indicate, mobile workers rely strongly on the ability to communicate synchronously with others. For the television production manager (and for others in our study), the mobile phone made it possible to efficiently clarify issues and negotiate with others, making it the technology of choice especially in urgent situations. Being able to talk anywhere, anytime allowed him the possibility to take action and reach closure on issues that he otherwise would have found difficult while on the road. This can be contrasted with having to leave messages via email and fax, leaving issues unresolved.

Another important aspect of the mobile phone was the coupling of the technology with the person. Without the mobile, making contact would also have been much more difficult for those trying to reach him. As he indicates, life on the road is unpredictable.

Finally, the mobile phone allowed an easy way for our participants simply to "touch base" with others and to keep informed of events going on while away – a form of remote background monitoring activity. Participants found it useful to phone the office "just in case" there was anything urgent and also to keep abreast of general issues that might impact their understanding of a situation and therefore their job. This was seen to be important not just for dealing with issues while away but also to help with the catch-up period on returning to the office. Much of this seemed to be entangled with social banter:

> ... we've always had a habit keeping ourselves, keeping one another up to date as the day goes on ... if he's down in London I would say, you know, OK we've had a brilliant day or we've had a bad day, you know we've just got a habit of doing that so ... I suppose sometimes it's just social banter, you know he'd say he had a good day, he had a bad day that type of thing, and by the way I bumped into so and so, do you remember him, you know we saw his project a year ago when it crashed.

The mobile phone was also an important asynchronous communication tool in terms of access to voice mail. With voice mail it was possible to check for new messages at more regular intervals during the day than was possible with email. Participants could more easily check it during a spare five minutes here and there and were much less restricted in terms of the geographical constraints on where it could be done. As such it was often used as the preferred channel for more urgent asynchronous communications.

12.5.2 Email

In terms of frequency of incoming communications approximately two thirds of incoming communication was via email. However, this was not distributed evenly across all the subjects and was accounted for by only five of the participants who used email facilities while mobile (receiving between 30 and 40 emails in the diary days we focused on). In addition, this figure also includes many email messages that constituted junk mail or messages that were relatively unimportant to the receiver.

Email was used less for outgoing than it was for incoming communication and was not used for managing urgent situations in the way that the mobile phone was. For the small group of participants who did use email, it was generally something that was done only when they had sufficient time available, making it less oppor-tunistic and reactive in its use than the phone. Email was only used by those par-ticipants who were involved in overnight stay situations. The typical scenario of its use was in the evening at a hotel, for the purpose of keeping on top of a build-up of messages that would otherwise have to be dealt with on their return to the office – a case of more evenly distributing their workload over time.

12.5.3 Fax

Fax, while an important communication device for some of the participants in office-based work, fax did not play a significant role for incoming communication while mobile (only two incoming faxes were received across the sample as a whole). It was regarded by some of the participants more as a back-up mechanism in case, for example, they had left an important document behind. As with incom-ing communications, the fax was used relatively infrequently for outgoing com-munications (seven outgoing faxes across the sample as a whole). When it was used, it was typically for the purpose of conveying information that was difficult to accomplish verbally, for example information that the receiver would need to read and reflect on. It was also a means of avoiding time-consuming conversation that might take place with a phone call.

12.5.4 Document Use Activities

Another kind of activity we looked at in more detail was our subjects' use of doc-uments, both paper and electronic. Most of these workers' activities were centred on use of paper documents. This was true of the documents these workers gath-ered to take away with them, the ones they used during travel, and the documents they brought back with them.

For example, whilst a few of the participants took electronic documents with them (e.g. a PowerPoint presentation or a work schedule), the documents they assembled prior to travel were typically in paper format. These included things like printed agendas, faxes, printed emails, printed reference and discussion docu-ments "just in case" they were needed, printouts of documents to read during free time, and working customer information documents.

The reasons for taking paper was that paper has important characteristics that make it useful for opportunistic appropriation in a range of circumstances both predictable and unpredictable that characterise mobile work. In mobile meeting sit-uations, small amounts of paper were portable enough to be taken to the meeting, immediately viewable for ad hoc reference and referral as required and afforded a high level of micro-mobility (Luff and Heath, 1998) around the meeting space that made it a useful conversational resource. Likewise, paper was frequently used for ad hoc reading activities, as illustrated by the number of participants who printed doc-uments out or carried paper documents specifically to read if they got any spare time and had to work in places where they could not do other types of work (because of a lack of resources). Our equipment salesman was a prime example of this. He would frequently use free time between meetings to deal with a briefcase full of paper customer files that provided him with quick access to (trip unrelated) information while on the phone to clients. The use of paper was ecologically flexible and the participants knew that it could be accessed opportunistically in contexts where other technologies would be awkward to use. This ecological flexibility meant that carrying paper was perceived as a reliable option. Paper use was not hindered by the complications of technology infrastructure incompatibility or breakdown. As such it was sometimes carried as a backup to electronic documents.

For those who took laptops, access to electronic documents was possible. Only three participants took their laptops to the meetings. Use during these meetings was primarily for the purposes of presenting slides rather than the more ad hoc reference to unanticipated documents. One of the three was actually catching up with some previously downloaded email during a particularly quiet part of a meeting. There was very little in the way of new electronic document creation (aside from email) seen in the diary episodes, though there was some minor editing of presentation slides before meetings. In support of spare/dead time activity, documents on laptops were not used except for more lengthy periods of dead time and were not typically used for dead time outside of hotel rooms. Only one of the participants actually used his laptop at the airport during dead time while waiting for a delayed plane – this again was for reviewing email previously downloaded in the hotel room. In this respect, electronic document activity with laptops can be regarded as less casual and opportunistic than paper document usage. This is perhaps somewhat paradoxical given that one of the benefits of taking a laptop is access to documents whose need was unanticipated prior to going on the trip.

In addition to what they took with them and what they used, these business travellers also gathered documents. Again, these were mainly paper documents. While they occasionally requested electronic documents, delivery of these documents was generally deferred until their return to the office. Likewise, email received during the course of travel were generally not for the immediate purposes of the travel episode. Conversely, gathered paper-based documents such as promotional materials, handouts, agendas and discussion documents were deliberately prepared for the purposes of the meeting, and to be used *during* the meeting. Paper was recognised as the best medium in support of face-to-face talk, a topic that has been written on extensively by others (Lamming et al., 2000).

Document distribution was also a feature of activity while travelling, although less so than gathering, with only half as many documents distributed as received. Distribution was by email (5490), hand delivery (3990), or fax (790). While email was highest, its use was much more variable than hand delivery. Unlike email, hand delivery was used to support the immediate needs of the situation and the work to be done around the document.

The final important document-related activity to mention is note-making. All of our participants at some point made notes or annotated documents. Both took place both within meetings and during phone calls, to record action items, clarifications, contact details and discussion issues, as well as reflective reading during free time.

While the study generated a huge corpus of data, one finding which was particularly striking was the extent to which talk in a mobile context was supported by documents, and also vice versa: document use was very often in conjunction with talk, whether it be face-to-face talk or on the mobile phone. These document-phone relationships in particular illustrate the ways in which the study participants made use of available tools and artefacts to accomplish their goals. They also show ways in which their technological resources limited what they could do, making conversation around documents problematic.

Considering each episode where both documents and mobile phones were used together, we found that the collection of episodes could be split into two groups: *docucentric* and *telecentric* interactions. Docucentric interactions were those where a document was the primary focus of attention during the episode, whereas telecentric interactions were more focused on the talk as the means of accomplishing the work. We found five main forms of docucentric interaction and four main forms of telecentric interaction that we will briefly describe.

12.5.5 Docucentric Interactions

12.5.5.1 Documents Triggering Phone Calls

Phone calls were sometimes triggered by the caller reading a document. In certain cases, this behaviour was necessary to the sender for the purposes of clarification:

> Sometimes it would be much clearer with a phone call because sometimes you find you've done six emails back and forth when one phone call would have got it all.

In this case, the phone call allowed clarification to be achieved thorough dialogue in a way that would be cumbersome through just document exchange.

In other cases, these documents were messages or queries that required a verbal response from the person being called. Often these document-triggered conversations involved a great deal of back and forth activity:

> I have a blue book that I write, it's my book that I use, my bible basically that I write everything down on and then I speak to so and so over the phone, put a pencil on something, then I have to go back to the director, my producer and then things roll on from there and then usually I'll confirm it, write down on the fax all the details of what I need and then they'll send me a fax through for costs of exactly what I've got because I quote for everything, they send you a quote through. You might have to get back to them then and say this quotes too much and they come back and give you another quote and then finally everything's sorted.

12.5.5.2 Phone Calls to Confirm Delivery of Documents

The phone was sometimes also used to check whether a fax had been received and acted upon. This stems from the difficulty of ensuring both that the document (e.g. fax or email) had been accurately sent, and that the recipient had received it or been able to access its content. This knowledge is important to the sender: it is not enough to know that they have sent a document, but they must also know that it has been retrieved by the recipient. Document sending when making hotel bookings is a typical instance of this: callers need to know that their booking information has been received so that when they arrive at the hotel, they do not need to worry that they have somewhere to stay.

The mobile phone was also used to draw attention to the fact that a document had been sent and needed to be looked at urgently. This was particularly important for the mobile professional receiving incoming information. For example, given that email communication was not something that was checked on a frequent basis while away on travel[1] verbal communication over the mobile phone (including voice mail access) was a more efficient way of attracting attention to a sent document than an email message:

> The mode of work these days seems to be that urgent issues get communicated through voice mail. Email is less of a tool for urgent communication. Sometimes voice mail is sent referring to email. No details on the telephone but it needs urgent attention. Voice mail is a more immediate contact form today ... It's more accessible because you can do it from far more places geographically than you can email – car, airport lounges, home, as opposed to which email needs to be in the office most of the time. It's also quicker both to connect and to listen to and to respond to.

12.5.5.3 Phone Calls to Elaborate on Documents

Phone calls were sometimes used to build a context around the purpose of a document and any actions associated with it. Documents often did not refer directly to an activity, or were incomplete and required further explanation. This additional information could be easily conveyed over a telephone call. In the case of the mobile professionals, the mobile telephone was a critical technology for this purpose. The interview with the managing director of the Italian import company illustrates this:

> Often if it's a major fax, I'll call after the fax and go through it with them ... Check they've got it and check they're going to do something about it and check that they understood it fully.

The mobile phone was also used as a back-up technology, as in the case of problems with the fax with transmission quality. In the following instance, use of the phone and the fax provided a combined solution to the problem of following directions to a hotel, the telephone adding a degree of flexibility to communication:

> I was a bit concerned that she might get a bit lost so I said I'd leave my mobile on, ring me any time – because the fax I sent through to her, she didn't get the information on time. It was last minute, she needed a hotel, so I faxed it through.

12.5.5.4 Phone Calls to Access Remote Documents

In some cases, people made calls to access information on or about documents that they did not have direct access to whilst they were away from their home office base. In these instances, they would call up the owner of the document, or someone who had access to the document, and get them to read out or forward the information from that document to them. In one instance, an interviewee said he occasionally forgot key documents and was often asked for information that he did not have on him, and he asked his office to fax the document to his hotel's fax machine:

> The fax solves a lot of problems.

Another example of this behaviour involved a request from the brewery account manager for an electronic document to be emailed to her. A phone call on the mobile phone was necessary for this because email was something that she could only conveniently check outside office hours when back in the hotel room. The mobile phone allowed her to get closure on the task as and when it came to mind. In addition, the urgency of the situation required that immediate feedback be given that the task was in hand. This was something that was best achieved through the synchronous communication of the mobile telephone.

12.5.5.5 Phone Calls for Device Proxying

A related issue to the above was the use of the mobile phone as a device proxy. While on business travel, the mobile professionals often do not have convenient access to the same document technologies and resources that they have back in the office. These needs also arise opportunistically within the unfamiliar and unpredictable environments where they must work while on business. In these circumstances, the mobile phone served as an extremely versatile tool that acted as a proxy to document technologies such as fax machines back in the office. For example, the lab equipment salesman needed to send a fax to a customer while in his car. Because he did not have access to a fax in the car and also because he did not readily have access to the information in document form,[2] he made a call on his mobile phone to his office requesting that they find the necessary document and fax the full details to the customer.

Other examples of this include the use of the mobile phone to dictate letters and the use of the mobile phone to listen to email being read and to dictate responses:

> No but if it was urgent there would be enough information that I could ring up the office normally and speak to my secretary, she does shorthand and she can type it as quick as I can say over my mobile phone, you know letter to so and so really urgent must go out, dear Mr so and so reference our conversation I have pleasure in quoting you for this blah, blah, blah, that's the price Linda, you know and she'll end and whatever it and I'll say nip in my drawer and get the technical information, get it in the post this afternoon, he's really chasing it. You know that sort of thing happens but I can do that on the phone. I can do most things verbally.

Whilst the telephone does act as a device proxy, it is not always the ideal task for the job:

> I mean I use a mobile telephone probably because I haven't got a mobile fax, it would be nicer for me because I find myself ringing up one person in the office and

I might ring him up ten minutes later to say I'd forgotten something so if I could sort of you know during the day jot down everything I had to tell one person in the office and send off a fax then you've also got the written record and so it's safer.

12.5.6 Telecentric Interactions

12.5.6.1 Document Discussion During Phone Calls

As with face-to-face conversations, a number of the telephone calls were based around some form of document discussion. These would range from a simple quick glance to reference information in support of the conversations to more in-depth document discussion that could also involve some form of annotation of the discussion document. Paper documents were particularly important in this activity because of their viewing and annotation properties. Also, like the phone, paper could be used flexibly within the wide range of ecological circumstances encountered by the mobile professional without being affected by the technology infrastructure constraints of locations such as meeting rooms, hotel rooms or in cars.

The importance of documents as conversational resources can be seen through the difficulties experienced by mobile professionals in certain situations. For example, one participant expressed some of the difficulties when not being able to view documents in mobile situations such as in the car:

> You do need to see the information, if I could see it myself it would be a lot easier ... I'm asking so many questions, is it this, is it that, can you see this, can you see that?

12.5.6.2 Note-taking During Phone Calls

People frequently needed to make notes when they were telephoning, for recording either contact details, action items or information that needed to be discussed there and then. Often such note-making made use of whatever paper was to hand in order to avoid disrupting the conversation:

> Did I make any notes, yes I made some notes on the newspaper because she called me so I made it on the back of a newspaper, two points ... just so that I could again capture the thoughts as she was going through it ... the numbers that I had in there I would be able to play with them and look down at them and reference them

Although the act of note-making while on the phone is not exactly a novel finding per se, these activities nevertheless presented some particular problems for mobile professionals because of their limited resources for writing or scribbling. Within the car this characteristic of telephone behaviour would sometimes cause difficulties for those actually on the move such as drivers, who regularly made and received calls whilst on the move:

> No it is a problem when you're driving, the mobile phone and the messages and remembering things, writing things down. I haven't managed to solve that problem yet.

Some important information was retained in the technology, such as the number of the caller being retained by the mobile telephone. For other, and particularly for complex information, many people had to resort to scribbling notes down whilst they were driving, or pull over to the side of the road.

12.5.6.3 Documents to Elaborate on Phone Calls

In certain instances, verbal communication in itself was not sufficient to convey all the information necessary for the task situation. Additional follow-up material, such as fax, email or posted documents would sometimes need to be distributed to elaborate on the information in some way. In the cases of people faxing documents on the move, it followed on from a phone call in over 40% of occasions. For example, one of the participants was in his car where he had a free hour available so he made a phone call to a customer using his mobile about some equipment he was trying to sell. While he was able to give the customer enough overview information to get them interested, they wanted full written details to peruse before they would commit to buying. The phone call was not sufficient in itself to make the sale because the customer needed time to look through the details more closely and in his own time. There was therefore a need to support the verbal telephone communication with paper-based visual information that would support the customer's needs. This required that some information be faxed to the customer in support of the phone call. Because he did not have access to a fax in the car and did not readily have access to the information in document form, he made a call on his mobile phone to his office requesting that they find the necessary document and fax the full details to the customer.

12.5.6.4 Documents as Records of Phone Calls

In some cases, people needed to have records of telephone calls. These were required for a number of reasons, including the recording of telephone numbers or client names for later archiving and retrieval. Sometimes they contained detailed information that was drawn from the telephone call to use later (an example being telemarketing surveys). Other follow-up documents confirmed the details of the call as an official record of the conversation. In the example below, the TV production manager was asked how he made a booking for film stock. He used a paper record so that he knew exactly what he had ordered:

> Just being on the phone basically, being on the phone, speaking to people over the phone and then putting something down on a fax to confirm it ... I've got to get everything down on paper so I know in my mind what I've got, what I've got coming, who I've got coming and then basically slap it in lists.

12.6 Implications for Mobile Technologies

The findings have highlighted an interesting interrelationship between mobile phone and document activities of mobile professional work practices that has hitherto been underplayed. Categorising this relationship between mobile communication and mobile document activities in this way provides us with a useful framework within which to think about new technologies. The categorisation helps highlight existing problems and can suggest new opportunities. Such a framework can also be used to provide a basis on which some initial assessments about emerging technologies can be made in terms of their role within the work practices of the mobile professional.

When thinking about the design implications of these findings it is important to consider the particular need of the mobile professional for technologies that can flexibly accommodate their information needs across the wide range of unpredictable circumstances and contexts. One of the reasons why artefacts such as the mobile phone and paper were so useful to these people is precisely because they respect this need. They offer "lightweight" solutions that allow creative use on the fly rather than trying to predict all problems and throwing technology at each and every one. Design implications should leverage these artefacts and build upon existing widespread technology infrastructures.

Bearing these issues in mind let us consider some potential technologies that are suggested by the taxonomy. A logical starting point for us to consider would be a potential relationship between scanning technologies and mobile phones. Scanning technologies integrated with mobile phones might offer a number of opportunities to integrate phone and document use. For example, in the case of *documents triggering phone calls*, small scan heads within a mobile phone could be used to access contact information from document cover pages. By tethering the scanner to the phone, software could even convert these marks on paper into phone numbers that could be immediately dialled at the press of a button. Furthermore, replying to queries in particular sections of text might be made easier by allowing callers to scan in the relevant sections and send them on for the call recipient to look at during the phone call.

Taking this further, larger scan heads might be used for the purposes of scanning in whole documents. This would help in the sharing and clarifying of documents as we discussed in Section 12.5.6.3. It would also allow for quick follow-up to phone calls for notes and records taken during a conversation (i.e. as in Section 12.5.6.4). Such solutions could be based around integrating scan head technology within mobile phones or around tethering portable scanning appliances with mobile phone technology. These can then allow document distribution through fax and internet channels.

There are also implications around technologies that more closely integrate access and distribution of electronic documents with mobile phone technology. Web-based document repositories may provide some benefit here by offering widespread accessibility to documents. Xerox's MobileDoc system (formerly the Satchel system) (Lamming *et al.*, 2000) is designed to confer these sorts of benefits by allowing remote access to electronic document repositories through the internet using simple document "tokens". Tokens can be beamed to "Satchel-enabled" devices for printing or viewing. These kinds of technologies are important because they leverage the ability of the mobile phone to access documents, which can then either be printed out or viewed on laptops. Mobile workers can then more easily send documents promised in conversations, or even jointly discuss them while talking. However, the success of these activities depends to some extent on the mobile worker's surrounding infrastructure, the ability to connect to the web, for example, or the availability of Satchel-enabled printers or laptop displays. In many respects it is good that such systems employ existing technology infrastructures that can be exploited when available. But as we have seen, part of what the mobile worker needs is freedom to work without infrastructure constraints. Thus their dependence on these additional technologies may mitigate their value.

Other technology options to be considered within this framework are shared displays for the purposes of discussing documents during phone calls. This kind of approach has been explored extensively in the CSCW literature (e.g. Ishii, 1992). But it has not really been used extensively within a mobile situation. This may be an opportunity to build upon technologies that support simultaneous voice and data transmission to allow remote people to talk over shared documents; for example, developing a version of HP's Omnishare for a mobile situation using a combination of Omnishare software, laptop with touchscreen and mobile phone. Cameras on laptops or mobile phones may offer some solution in this area that deserves some consideration and investigation, though as with the shared display technologies, they may be ultimately too cumbersome for these particular circumstances.

The audio facilities of mobile phones could also be exploited for various categories within the taxonomy. Recording audio snippets such as contact details and action items during a phone call could support activities within the "note-making during phone calls" category especially for in-car conversations where other forms of note-taking are difficult. The legal implications of this are perhaps preventative but other audio recording facilities could be integrated within the phone for after-call recording of action items and contact details. Such audio recording facilities might also be used in support of activities within the "phone calls to elaborate on documents" category, providing context in the form of a short verbal message that can be attached to documents both paper and electronic. For example, audio annotations could be created and played back using mobile phone technology. Audio files could be attached to emails or linked with paper documents and accessed through a URL or barcodes link.

Finally, techniques for viewing and annotating documents using PDAs while using the mobile phone provide interesting possibilities. For example, by tethering a mobile phone to a PDA or by incorporating PDA functionality into a phone, note-making during phone calls could be supported. Callers could make notes which are then automatically tagged with information about the phone call, such as details about where, when and who was called. This could provide records of phone calls. Callers can also make notes about action items that are then sent to their "to-do" list. This could be augmented by having an automatic "to-do" button that records a snippet of the conversation as a reminder to future action.

12.7 Conclusion

In conclusion, the findings have shown the importance of the mobile phone for the work of mobile professionals. In contrast to the laptop, the flexibility, versatility and convenience of the mobile phone have made it a ubiquitous device in terms of who owns one, whether it is taken on trips and where it is subsequently used. As such it is rather like paper, and the link between paper documents, and indeed documents in general, provides an important leverage point for thinking about new technology ideas. While providing some descriptive analysis of how mobile workers use both communication technology and documents to manage information on the move, we hope to have demonstrated that looking at the relationship between talk and text offers new insights for mobile technologies.

Acknowledgements

The authors would like to thank the study participants for generously giving their time and effort in this study. Thanks also to Marge Eldridge and Colin I'Anson for their comments on earlier drafts of the chapter.

Notes

1. For those who used email while away, it was generally something that they would check only once or maybe twice a day when they had opportunity to download it. The end of the day in the hotel room was a typical scenario for downloading and dealing with email.
2. This was interesting in itself that the participant did not have access to the information necessary to complete the demands of the phone call. Predicting the need for taking the document was not possible because of the whimsical nature of the phone call during what would otherwise be dead time in the car. As such, not only was the technology not available to make the fax but nor was the document. So while a mobile faxing device could have been useful in situations like these, this must be interpreted within the context that mobile professionals will often not have the information they want to be able to make the fax.

References

Ablondi WF and Elliott TR (1993) Mobile professional segmentation study. *BIS Strategic Decisions report*.

Eldridge M, Lamming M, Flynn M, Jones C and Pendlebury D (1999) Research methods used to support development of Satchel. *Proceedings of INTERACT 1999*, Edinburgh, Scotland.

Ishii H (1992) Clearboard: A seamless medium for shared drawing and conversation with eye contact. *Proceedings of CHI '92*, ACM Press, pp. 525–532.

Lamming M, Eldridge M, Flynn M, Jones C and Pendlebury D (2000) Satchel: Providing access to any document, any time, anywhere. *Transactions on computer–human interaction, special Issues entitled Beyond the workstation: human interaction with mobile systems*, 7 (3), pp. 322–352.

Luff P and Heath C (1998) Mobility in collaboration. In *Proceedings of CSCW '98*, ACM Press, pp. 305–314.

Puuronen S and Savolainen V (1997) Mobile information systems – an executive's view. *Information Systems Journal*, 7, pp. 3–20.

Sellen A and Harper R (1997) Paper as an analytic resource for the design of new technologies. In *Proceedings of CHI 97: Conference on Human Factors In Computing Systems*. New York: ACM Press, pp. 131–137.

Whittaker S, Frohlich D and Daly-Jones O (1994) Informal workplace communication: What is it like and how might we support it? In *Proceedings of CHI 94: Conference on Human Factors In Computing Systems*. New York: ACM Press, pp. 131–137.

Chapter 13
Usability of Portable Devices: The Case of WAP

Vincent Helyar, Serco Usability Services, London, England

13.1 Introduction

> Imagine stepping out of an office building on the way to the airport and using
> your WAP-enabled, wireless device to check the traffic report. Finding congestion,
> you locate the train timetable and choose to purchase a train ticket on-line instead
> of driving. On the way to the airport, you select your aisle seat, check in for the
> flight, and reserve a special meal. Finally, you unpack your raincoat after looking
> up the weather at your destination (WAP Forum, 1999).

Mobile phones are the best selling communication devices of all time. In the UK
about one in three people own a mobile, while in Finland penetration is measured
at almost 70%. Another growth area, the internet, is seen as being closely related to
the mobile market. It is predicted that all of the one billion mobile users expected
in 2005 will also be connected to the internet, by which point convergence will
perhaps have dissolved the difference between the two services (Ericsson Radio
Systems, 1999).

WAP (wireless application protocol) is "the de facto standard for providing
Internet communications and advanced telephony services on digital mobile
phones, pagers, personal digital assistants and other wireless terminals"
(www.wapforum.org). Using WAP, content can be delivered over the internet to
most current wireless networks and some new networks including GPRS (General
Packet Radio System) and 3G (Universal Mobile Telecommunications System).

WAP is designed specifically for the mobile market using existing internet standards (such as XML and IP), and has been tailored to cope with the constraints of
the mobile environment such as intermittent coverage, low bandwidth and one-
handed navigation styles. The development language often used for WAP, Wireless
Mark Up Language, aims to make optimum use of the small screens found on WAP
phones and incorporates scalability to deliver content in the most attractive way
possible to devices with larger screens such as personal digital assistants.

When WAP became widely available in the UK at the beginning of 2000 it was
received with scepticism by the media, and disappointment and rejection from
users. This chapter reports the findings of an independent usability study carried

out on WAP services that were available at its first launch in January 2000. The chapter also discusses the implications of these early findings with respect to recent developments and emerging technologies.

13.2 Usability

Numerous constraints on the technology and user interface of mobile devices make usability a vital consideration. At the moment mobile devices suffer from small screens, poor input methods, low bandwidth and limited battery life. Services must be carefully designed to fulfil users' needs, without overloading them with unnecessary complexity or slow operation. This requires very careful user needs analysis and elegant design. At the moment, WAP sites in the UK are largely competing for hits; however with the growth of m-commerce, these sites may compete for revenue in the future – be it in the form of actual purchases made via mobile phone, or as monthly subscriptions and micro-payments as in Japan.

While at the moment WAP users are to a certain extent kept within the boundaries of their portal pages, both by their provider and by the limited number of WAP sites available, this will not always be the case. As WAP becomes more popular many sites will inevitably suffer from bad design, and will contribute to usability problems for the technology as a whole. In this context usability will become crucially important to the success of WAP services. The "cost" of switching between WAP sites will be lowered, and before long users will be able to take their query or purchase to a competing WAP service with minimal effort.

As with the web, switching costs on WAP are low and will only get lower. "Switching costs" refers to the amount of effort that it takes a consumer to switch from one retailer to another. In the physical world, switching costs are high for consumers – people tend to feel committed to their choice of shop even when the service is bad, and especially when required to make significant effort to reach the vendor in the first place. On the web the amount of effort required to reach competitor vendors is minimal, only a click away. Research of user behaviour on the internet has repeatedly demonstrated a low tolerance of designs that are difficult to use and sites that are slow to load (Nielsen and Norman, 2000).

If properly introduced into the product development lifecycle, usability can ensure that WAP services and WAP content are both useful and usable. WAP services must be designed to add value to the user experience, deliver the right kind of content at the right time and in the right place. Usability studies, such as the one described in this chapter, go some way to identifying the nature of content that users would use on their mobile handset, and highlight some of the usability pitfalls that can stop the willing user from making effective use of a WAP service.

13.3 User Perceptions of a First Generation WAP Service

In January 2000, Serco Usability Services carried out a WAP usability study in collaboration with a major UK network provider. The purpose of the evaluation was to gauge the usability of WAP services using the Nokia 7110 and the Orange portal, the only WAP phone and service available at the time. The findings of the study have been circulated widely in the form of industry guidelines on designing usable WAP services. The following sections of this chapter describe the findings of the study; recommendations and the implications of the findings.

13.3.1 Method

This study used qualitative techniques to gather data from a total of 12 users. Usability testing experience with the web has consistently shown that most key usability problems will be revealed even with a small sample; Nielsen estimates that 80% of "site level" (i.e. issues not related to individual pages) problems will be uncovered using just five users (Nielsen, 1998). It was thought appropriate to use a larger sample in this instance to allow for a greater degree of flexibility during the trials, given that the focus of the trials was technology new to both the researchers and users.

The 12 users were recruited by a specialist agency; users had a range of internet experience and experience with mobile phones:

- 10 used the internet regularly
- 10 owned a mobile phone
- three had shopped online before
- 11 were aged 30 or less.

The evaluation took place in a purpose built usability laboratory, consisting of a test room and an observation room, separated by a one-way mirror. Observers were able to view the same content as the user, by means of a ceiling-mounted camera zoomed in to the screen of the phone. The co-operative evaluation involved users interacting with the WAP phone, while accompanied by a trained facilitator. The role of the facilitator in the exercise was to use neutral questioning where necessary, to prompt for feedback on the service in use.

At the beginning of each testing session users were introduced to the structure of the evaluation, although they were not told at first that WAP content was the main area of interest. Following this introduction, users were asked to imagine that they had just bought this new phone and to demonstrate what they would do with it.

The course of the testing session was as follows.

13.3.1.1 Exploration Phase

Users were encouraged to look at content of specific interest to them; the facilitator did not draw their attention to anything during this phase. The purpose of this phase was to gain an understanding of what areas of the phone were interesting to users, and to observe whether users showed any natural interest in WAP content.

13.3.1.2 Task Phase

The facilitator asked the user to find specific pieces of information (e.g. cinema listings, a business address) using the WAP phone and the portal provided. The purpose of this phase was to gain an understanding of whether users were able to use the WAP service provided to find specific pieces of information.

At the end of the evaluation, users were asked for final comments on the service they had used and issues of particular interest raised during the course of the test were revisited where necessary.

13.4 Results

This section details the key findings of the usability evaluation carried out, but does not cover the full range of issues raised by the research. The chapter then considers the implications of the findings, with respect to current and future technologies.

13.4.1 Navigation and Terminology

Navigation conventions on the phone that was tested were drawn from both computers and mobile phones. These inconsistent navigation structures resulted in many users building a confused mental model, such that the "back" key appeared to work in different ways depending on where in the phone's menu system the user was. For example, in areas of the 7110 other than WAP content (such as the address book or settings screens) "back" operated by taking the user up one level in the menu hierarchy (as on most Nokia phones). In WAP content "back" appeared to behave like an internet "back", taking the user to the last page seen, and not necessarily ascending the menu hierarchy. The confused mental model that users built meant that it was difficult for them to grasp what the "back" option was actually doing when in WAP services, and meant that the phone frequently behaved in an unexpected way. Participants in this study expected "back" to return them up a level in the menu hierarchy, rather than take them back historically to the previous page.

Users had difficulty understanding computer terminology used, such as home, bookmark, template, empty cache and security certificate – even computer literate users did not always understand such terms. Home was frequently associated with a home phone number in this context, and three users cancelled the call on initial connection to the WAP service fearing that they were calling a home phone number. The prominence of the number called by the phone to access WAP content appeared to raise concerns in users, in a way dialling information when using a modem on a PC would not, "... they do give you what they're calling [on a PC] at the bottom, but I don't take much notice of it". This evidence suggests that computer metaphors may not always transfer smoothly to new devices which are neither phones nor computers. Other operators involved in mobile technology

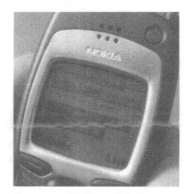

Figure 13.1 Confusing metaphors and terminology – what does "back" actually do?

Figure 13.2 The participant wants to select a region from a list – they are unsure what button to press, but finally decide on "Options".

have been keen to distance their products from computers, so as not to discourage potential customers who are not internet literate:

> NTT DoCoMo has never used the word "Internet" in the marketing of its service ... it is presented as a group of new mobile services, enabling the operator to attract a high proportion of non-Internet users. (Bond, 2000)

As part of the task phase of the evaluation, users were asked to locate the address details of Imperial College in London using the business directory service available on the Orange portal site at the time. Of the 12 participants in the evaluation only two were able to complete the task, and it took them over 25 minutes of searching to find what they were looking for.

The principal reason for users' failure at this task was that the information architecture of this directory service was such that it forced participants to "browse" through multiple categories to seek information. The browse model that was used on this service did not help users to find what they were looking for – they were presented with ambiguous terminology for categories, and were required to drill fairly deeply in to the service before they received any feedback about the success or failure of their enquiry. The majority of users that took part in this evaluation spent a long time mining for information only to find that they had started off in the wrong category. Those that did manage to retrieve a list of universities in central London had to sift through a list of up to 132 results, displayed four lines at a time.

The architecture of this WAP service is a good example of what can result if users are not involved in the development process. In this instance, users wanted to be able to find information quickly, before backing out of the service to focus their attention elsewhere. Concerns about cost, the nature of the mobile environment and the bland presentation of information on the screen all contributed to users wanting to "seek and retreat" – users in this trial were not interested in browsing WAP content. Indeed, it is difficult to imagine when content displayed on a device with such a small screen would be suited to a browsing metaphor:

> On the Internet, the user is quite happy to browse until they find what they want, sometimes they are happy to browse for no other reason. Behaviour of mobile users is more impulsive – they have no time to browse through pages of content, they are seeking specific information now and they need it at their fingertips. (Logica, 2000)

For this reason it is reasonable to expect that pages of content that are relevant to being bookmarked (e.g. the results page of your local cinema) are likely to be revisited by WAP users more frequently. Bookmarking is currently the most obvious way of keeping browsing to a minimum, and maximising the value for money of content. More recent usability evaluations of WAP services have confirmed the importance of bookmarking in the usability of WAP services (Ramsay and Nielsen, 2000).

13.5 The Right Approach to Portability

In October 1999, WAP was dubbed the "Wrong Approach to Portability" (Nielsen, 1999). Nielsen claimed that "impoverished user interfaces" would result due to the lack of context on a small screen and poor input methods available on telephones. Nielsen's claims still stand today with respect to displaying "internet" content on a mobile phone; however this does not mean that there is no content that can be both useful and usable when displayed in this context. There is a "Right Approach to Portability".

This study demonstrates that content taken from one kind of media and transferred to another rarely works. This lesson has been learnt when transferring content from print to the web and then from the web to interactive television (Daly-Jones and Carey, 2000). Transferring content from the web to the mobile domain is no easier. For content to be useful and usable in the mobile environment, it must be designed for the mobile environment; likewise for content to work on the web and interactive TV, it must be designed for those contexts. The results of this study suggest that there are a number of critical issues that should be considered when designing for the mobile context.

WAP content must be tuned for *occasional and transactional use* – awkward navigation, small screens and unreliable connections do not encourage users to spend long periods using WAP services. User perceptions may change as billing structures alter, and bandwidth increases. Furthermore, with the arrival of a new generation of mobile devices, users may feel more comfortable with lengthy visual interactions with their handheld if the screen size and appearance lend themselves to this (see Figure 13.3).

Consumers need *quick closure* – users will not spend long periods looking for the information that they are after, and importantly, WAP is rarely likely to have the luxury of a user's uninterrupted attention. There are many other forms of media available to the user (e.g. newspapers), which are often more cheaply and easily available. WAP services must give the user confidence that they can find the information they are after as quickly and easily as competing media.

Personalised content will be popular with users – input of information to WAP is painful, and the mobile phone is a personal object. However, experience with the

Figure 13.3 A Nokia 3G terminal concept
(Reprinted by kind permission of Nokia © 2000)

web shows that users are unlikely to devote large amounts of time to personalisation unless it is obvious that it is of real benefit to them.

WAP services must *add value*. Users are currently suffering from an information overload, from the web and from print media. For WAP to gain a foothold in this crowded market place it must add value to the user's experience, in offering a kind of service which is not available elsewhere. Location based services will offer the user a service that has not before been easily available. Such services capitalise on the mobile context, and add value to the user's experience of accessing information on the move.

Currently there are few real indications of what services may be appropriate for WAP and other content-based mobile services. However, a simple framework that will allow the identification of promising products can be sketched out. Such a framework may include indicators of benefit (often unique to the mobile context) such as the degree of availability, the currency of information required, the opportunity for immediate action and other attributes illustrated in Figure 13.4. These benefits can be mapped against potential mobile internet services, to illustrate the degree to which a service may reflect the benefit. The greater number of benefit indicators (e.g. "always available", "current", etc.) that a potential service can satisfy, the closer that service comes to capitalising on the mobile context. Figure 13.4, the hypothesised "Power Awareness Reach Context" model, is essentially a means for determining the degree to which a service present on the mobile internet will *add value* to the user experience of that service. The indicators of benefit presented in the first column of the table, do not make an exhaustive list and may vary depending on the capabilities of the mobile device and the idiosyncrasies of its user.

Figure 13.4 applies some examples of indicators of benefit, against some possible and current applications. The applications which are illustrated are a financial management service, a house purchase service, a "hitchhiker's guide" application which provides an intelligent portal, food shopping and a tool for choosing and buying a pet. The table indicates the ways in which these benefits may map to the applications.

Apart from the paucity of WAP services that add value to the user experience, the WAP standard itself is held back by the GSM network on which it operates in Europe. This is most evident in the billing structure that is currently in operation for WAP services in the UK, where the user is charged according to the "air time"

BENEFITS	P R O D U C T				
	Manage your finances	Buy a house	Hitchhiker's guide	Shop for food	Choose a pet online
Always available	+	+	+	+	–
Current	+	+	+	+	–
Anywhere	+	+	+	+	–
Immediate	+	+	+	+	–
Supports decisions	++	+	–	+	++
Allows control	+	–	–	+	–
Private	++	–	–	–	–
Personal	++	+	–	+	+
Comparative	–	+	+	++	+
Alerts	+	++	+	–	–
Restricted Content	+	–	–	+/–	++
Limited Interaction	+	–	+	+/–	–

++ Very relevant; + Relevant; - Not relevant

Figure 13.4 The PARC (Power Awareness Reach Context) model.

that he uses. This means the amount of time that he is connected to a WAP service, regardless of the amount of content that he receives or the speed at which he receives it. Users are effectively penalised for taking a long time to find the information that they want. The alternative is to charge the user for the amount of information received; this will become possible with the advent of packet switched networks (such as GPRS). This billing model has already proved successful with I-mode in Japan, providing a constant revenue stream from the mobile internet for the network operator. I-mode content providers take revenue from users by charging a small subscription fee to users; this also encourages the content provider to deliver a high quality product, in a bid to attract new subscribers.

13.5.1 Recent Developments

Following the soft launch of WAP in the UK in 1999, WAP services were aggressively marketed to consumers the following year. Heavy marketing campaigns have touted WAP as the "mobile internet". While this has become a popular term for referring to third party content available on mobile terminals, it is not an accurate description of WAP content. These insensitive campaigns may be held at least partly responsible for the disappointment felt by the early adopters of WAP, and for the subsequent "WAP backlash" covered in much of the European media:

> WAP is the biggest disappointment to hit the mobile phone world ... [when] WAP
> was described to journalists it was going to be "the Internet on your mobile

phone". The reality is altogether different. Not only is WAP incredibly slow, but instead of the limitless universe of the Web, WAP phones can only access special WAP pages, which are few and far between. (Homer, 2000)

Even with these events, WAP has grown considerably in the last year. There are now approximately 8 million WAP subscribers worldwide, and in the UK BT Cellnet (the second most popular network operator) has over 1 million WAP users – 10% of its total subscriber base. According to a recent press release, Genie (BT Cellnet's WAP portal) received 62.5 million page impressions during December 2000, a massive 513% increase over the quarter (BT Cellnet, 2000). However, page impressions are not a reliable indicator of popularity, as they say nothing of the number of "unique" visitors to the site. To the cynic, these phenomenal figures logged by Genie could just as well indicate difficulty experienced by users in navigating or accessing content.

The number of WAP devices available to the user has also grown significantly since the time the study reported in this chapter was carried out. Users can now choose from a proliferation of products with large screens (e.g. Ericsson R380), alternate navigation mechanisms (e.g. Sony CMDZ5) and browser types (e.g. openware, Nokia, Microsoft, etc.).

Competition is expected to appear in Europe in the form of I-mode, the highly popular Japanese "mobile internet" devised by mobile phone giant NTT DoCoMo. However, many WAP experts are sceptical of the potential for I-mode in Europe. NTT does not have a grip over Europe as it does over Japan. In Japan NTT has approximately 28 million subscribers, an incredible 20 million more than its closest competitor J-Phone (which ironically has adopted WAP). If NTT launches I-mode in Europe, it may face fierce competition from handset manufacturers, WAP founders and established WAP portals.

13.5.2 I-mode and 3G

Some industry experts believe that the user experience that will be offered by I-mode will be superior to WAP (Ramsay and Nielsen, 2000). I-mode operates on a packet-switched network so that phones are always connected to the network. Whereas WAP users must wait for their phones to dial up before they can begin to access services, I-mode users have immediate access to services through a dedicated button on their handsets. Packet-switched technology also allows DoCoMo to charge users according to the amount of information they download, rather than the time they spend online. This means that users are not charged while they are reading and writing messages, or waiting for a site to download. The billing model of WAP over GSM, coupled with difficult navigation, means that WAP is currently an expensive tool – yet another barrier to users (Ramsay and Nielsen, 2000).

Simple personalisation is supported on I-mode; within DoCoMo's portal, users can select which services they want to appear in "My menu", effectively an online bookmark service. Tedious log-on procedures where users must enter a username and password do not exist on I-mode. The system as a whole is also more reliable;

Figure 13.5 The D209i i-mode mobile phone.
(Reprinted by kind permission of NTT DoCoMo, Inc. © All rights reserved)

the frequent problems that WAP users experience when trying to access services are far less common on I-mode. However, many I-mode sites suffer from poor navigation in a similar way to many WAP sites (Svanteson, 2000).

Many of the key differences between WAP and I-mode are in the breadth and quality of the services on offer rather than their usability per se. DoCoMo worked proactively with content providers to ensure a good range of services when I-mode was launched. Great similarity between CHTML (the development language used for I-mode) and HTML has made it cheap and easy for developers to create new I-mode sites:

> NTT DoCoMo has ensured a large number of services within I-mode by creating low barriers to entry by merchants and users, setting up a virtuous cycle in which more content attracts more users and more users attract more content. (Bond, 2000)

The evolution towards third generation (3G) mobile technology promises ever increasing bandwidths, bringing with it new capabilities such as streaming audio and video content. However, all of this potential does not necessarily convert to an improved user experience. The real potential of 3G is not yet clear; however, the danger is that development will be driven by technology, not by user requirements. Such development does not bode well for the user experience.

3G devices will have larger screens, and may support pen input and interaction by direct manipulation. Developments like this will give content providers a much greater degree of flexibility to design a useful and usable service. These 3G devices will quite likely bear no resemblance to mobile phones, being geared to visual interaction as opposed to being designed for voice interaction. Nielsen predicts that it is precisely this kind of innovation that will need to happen for the mobile internet to gain the kind of popularity that is frequently cited in market projections (Nielsen, 2000).

13.6 Conclusion

The wireless application protocol is the first step in the mobile internet revolution. It has been designed as an open standard, which strives for compatibility across a range of different networks accessed by users with a variety of handsets. Accessing WAP over GSM, most common in Europe currently, means slow download speeds, simple text based services and unsuitable billing models. The "mobile internet" in the year 2000 proved to be hype more than anything else, disappointing both the media and consumers.

This chapter began by reporting the findings of a usability study carried out soon after the launch of WAP in the UK. The results of this study have been used to identify issues that are potentially critical to adopting the "right approach to portability". Services placed on the "mobile internet" must be designed for that context, and must exploit the characteristics of the context, often unique to being on the move. For example, a successful WAP service may be one that its user has a need to access anywhere and at any time. There are many of these "benefits" that may determine which services are successful, and which are not. They range from those that are inseparable from being mobile (e.g. access anywhere, at any time), to those that are dictated by technology to a certain extent (e.g. interaction is difficult and limited). The "Power Awareness Reach Context" model, illustrated in Figure 13.4, introduces a possible approach in identifying the potential of a mobile application.

As this chapter has illustrated, usability issues will be especially relevant in the development of WAP and emerging "mobile internet" services. Usability issues that exist at the moment exist largely not because the underlying protocol is at fault, but because those adopting the standard are not sufficiently user focused. It is the responsibility of WAP service designers and developers to make proper use of the standard to meet the requirements of users – do consumers really need another way to access news on the move? Having said this, the user experience is also affected by technology. The need to have a constant connection to access WAP raises the expense of using the service, and leaves users feeling frustrated when they spend money searching for content that they never find. WAP users in Europe will not benefit from the billing structure used by I-mode until the launch of packet switched networks, such as GPRS.

Usability research can help to ensure that lessons are learnt from current generations of mobile internet, which can then be applied to secure the success of the next generation of technology. Further research in this area must focus on the mobile devices themselves, the content displayed on them and the user interacting with them. Guidelines and design conventions resulting from such research will help to make accessible the widest possible amount of information. From this comes the range of choice, characteristic of the wired internet. Areas of particular interest for further research include identifying applications that will drive the mobile internet revolution, and identifying the consumer demographic to which such applications will appeal.

Acknowledgements

The author would like to thank Peter Thomas and Martin Colebourne for their help with this chapter.

References

Bond K (2000) A new market model on trial. *M-Commerce World*, November, 2000.

BT Cellnet (2001) New figures show mobile information and services are becoming a winner with consumers. Press release, January 15, 2001, http://www.btcellnet.co.uk

Daly-Jones and Carey (2000) Navigating your TV: The usability of electronic programme guides. *Proceedings of CHI'2000*, extended abstracts, The Hague, Netherlands, ACM Press.

Ericsson Radio Systems (1999) *Welcome to the third generation*. http://www.ericsson.com/

Homer S (2000) A new dawn: is the new 3G system a mobile phone revolution, or is it just the Emperor's new technology? *The Independent*, November 19, 2000.

Logica (2000) *The Mobile Internet Challenge*. Logica white paper (un-numbered).

Nielsen J (1998) Cost of user testing a website. Alertbox usability column, http://www.useit.com/alertbox/980503.html.

Nielsen J (1999) Graceful degradation of scalable Internet services. *Alertbox usability column*, http://www.useit.com/alertbox/991031.html.

Nielsen J (2000) WAP: miserable failure? Interview with Jacob Nielsen by Fred Johnson & Lorna Ho. *Yahoo! Vision*, http://vision.yahoo.com/search?sp=nielsen.

Nielsen J and Norman D (2000) Usability on the web isn't a luxury. *Information Week*, http://www.informationweek.com/773/web.htm.

Ramsay M and Nielsen J (2000) WAP Usability Déjà vu: 1994 All Over Again. *External usability report*, Nielsen Norman Group, http://www.nngroup.com/reports/wap/

Svanteson S (2000) *Particular challenges in designing for "baby faces"*. Masters thesis, Royal Institute of Technology, Stockholm.

WAP Forum (1999) What is WAP and WAP Forum? *FAQ page*, http://www.wapforum.org/faqs/index.htm.

Chapter 14

The Mobile Interface: Old Technologies and New Arguments

Richard Harper, Digital World Research Center, University of Surrey, England

14.1 Introduction

It is nearly two decades since Marvin wrote *When Old Technologies Were New* (Marvin, 1988). In that text, Marvin explored whether the impact of electronic communications at the end of the nineteenth century (and thereafter) turned out to be as predicted. For those who have not looked at that book, they may be unsurprised to learn that Marvin discovered that the pundits got it wrong. Instead of predicting that the greatest changes would occur in what goes on in home settings – through radio and television for example – they focused their predictions on the impact that new technologies would have on mass spectacles in public spaces.

In this chapter I want to reverse Marvin's thesis. Picking up on Brown's opening chapter to this book, I want to start by saying that mobile technologies aren't really new at all; indeed they have been around for many decades. (What is new is the fact that they are now readily available to the consumer, and the consumer has responded by buying them in vast numbers.) Given that this is so, I will then ask, are the arguments, explanations and analyses that have been represented in this book novel? In other words, the technology may be "old", but are the arguments new?

I will be deliberately critical here, not because the texts in question deserve criticism, but rather because I want to use this concluding chapter as a way of stirring up the arguments in the research domain. It seems to me that many researchers in this area – and I include not only those whose work is presented in this book but others too whose work is beginning to appear – are rather haunted by ideas from office information systems (OIS), human computer interaction (HCI) and, to a lesser extent, computer supported co-operative work (CSCW). Those papers that come from the sociological domain known as Social Studies of Science and Technology (SST) are particularly vulnerable to the claim that they don't explore new issues but rather revisit old ones.

Why would this matter? After all, just because an argument is an old one doesn't mean it is a bad one. Indeed some disciplines pride themselves on the fact that they don't really move on – take philosophy. Wittgenstein is said to have created a revolution in that discipline (at least twice!) and, though there can be no doubt that he put a new angle on philosophical topics, it is not unreasonable to say that he was dealing with the same old chestnuts that Kant fretted over two centuries before. But in this domain I think it is a problem, for at least two reasons. First, the ways in which mobile devices are used is not entirely as expected and indeed reflects changes in patterns of behaviour that force some revisiting of predictions about what the future may hold. For example and by way of illustration, we have been told (and indeed some of the terminal manufacturers continually assault us with the claim) that third generation networks will foster a multimedia future (see for example, Lundin and Persson 2001). Yet the ways that people are using mobile devices, particularly to support text messaging, seems to indicate that users are not too interested in the "data rich" future. They are perhaps more interested in revising the long mourned art of letter writing although perhaps Walpole and Gladstone (to name but two great letter writers) would have been appalled at the spelling and the brevity. So the future may not be as some are currently predicting; but what I would like to suggest is that how it may be different is discernible in the emerging patterns of behaviour one can see now.

So I am going to present a kind of literature review, albeit slanted with a concern for what empirical evidence is telling us about what the future may tell us. I will organise the review by locating the texts in question within a very simple schema, a schema that has to do with what one might call the "interface". It is the interface, after all, which is the all important glue between the network, the service provider, the phone manufacturer and the humble user. It seems to me that there are four ways in which the term interface can help us marshal and assess the literature on the mobile age. Each area treats the interface somewhat differently and each implies a different set of issues.

The first of these relates to what I will call the "current forms of the interface" on mobile phones. Here I will discuss the fact that there isn't that much to distinguish between the bulk of phones on the market: apart from shape and colour. Indeed the fixed line user from over a hundred years ago would probably manage to use them quite happily with a little bit of guidance. Several of the chapters in this book deal with this standard form, whether it be the sociologically informed chapters of Cooper and Green (Chapters 2 and 3) or the design oriented ones later on in the book, such as Helyar's (Chapter 13). My main concern will be to say that there is something of a settling of the form factor in the mobile domain around one particular design, and despite the hype about the future, this may result in the future being pretty much as we know it now. This is because there may be some resistance to moving away from this form factor.

Second, I will consider what I will call "radical form factors", by which I mean the possibility of breaking out of the mould of the current mobile phone towards so-called convergence technologies, where, for example, PDA functions are combined with communications devices. The chapters by O'Hara *et al.* (Chapter 12) and Churchill and Wakeford (Chapter 11) in this book are perhaps most obviously dealing with some of these issues, though I will also bring to bear some papers published elsewhere too.

I will then talk about what I will call "users' perceptions of the mobile interface", and here I will build on the chapter by Palen and Salzman in particular. I will address the fact that when someone uses a mobile device, what they interact with is not simply a piece of hardware nor yet the software on the device itself. They are also interacting with a network provider and a service agreement. These affect what they can do as well as what they think they can do. From this view, the interface is an amalgam, and consists of much more than is normally accepted within the domain of inquiry of HCI, for example, and is equally more complex and subtle than it is possible to address with traditional fieldwork-based observation techniques, as are often used in CSCW, for instance. The Palen and Salzman chapter is not the only one that I will consider here, since other chapters too deal with this issues, albeit not so directly. Laurier's description of the curious – and at times somewhat worrying – habits of a mobile professional highlight the fact that the use of mobile devices is always part of an assemblage of tasks and things.

And then I will deal with the fourth view on the interface, where I will address "interfaces to social processes". The general theme here will be to explore what we have learnt about the general relationship between the overall affordances of mobile interfaces and various patterns of social action. The chapters by Gant and Kiesler (Chapter 9), and Sherry and Salvador (Chapter 8) describe some of the ways in which mobile devices "fit into" the world of mobile professionals, for example. Perhaps more interestingly, however, the chapter by Weilenman and Larsson (Chapter 7) points out how the ways in which that fitting occurs in quite unexpected ways with teenagers. They don't use the technology to communicate with each other over distance for example, but when they are side by side; sometimes they even exchange the devices themselves.

All of this gives further credence to Cooper's claim in Chapter 2 that the mobile age should be forcing us to ask more questions about our expectations and theoretical models than we have to date. I will suggest that such unexpected practices point towards interesting ideas for the design of new services and devices. But to get there, one must focus on what current technological devices are actually used for. It is this that the technology pundits of the nineteenth century failed to do, focusing instead on their own hyperbole, and in turn making faulty predictions. If we do not attempt to understand what is already happening with mobile devices, we will misunderstand our own future. For those who might seek guidance as they make decisions about investments in third generation networks and equipment, this could be costly; for those in the business of exploring the present so as to grasp what the future may be, it is another demonstration of what a disappointing art it can sometimes be.

14.2 Current Forms of the Interface

In Europe, and in the UK particularly, there are now many shops which sell nothing else but mobile phones. This is a most curious development when one considers that no such development occurred for fixed line telephones. There are doubtless many reasons behind this, which will include the extraordinary uptake of mobile phones as consumer products, the fact that people like to change their devices frequently, as well as aggressive marketing. But what is particularly interesting – for the arguments concerning us anyway – is the fact that despite there often being hundreds of "terminals" on display in these shops, most of these terminals offer more or less the same "interface". They all have, necessarily, the 12 alphanumeric buttons of course; but most also have six so-called "soft buttons" . These allow the user to switch the device on or off, to scroll, to adjust volume, and so on. Each type of phone may use these six buttons for slightly different functions, but, as I say, nearly all have these six plus the essential 12 alphanumerics. There is then a basic interactional form for the mobile, arrayed around eighteen buttons.

Needless to say, some manufacturers do play around with this basic form but mostly this adds up to little more than alterations of colour and case design. There are also some general trends such as the move toward smaller size of phones. Some manufacturers have key form factor "differentiators" too, insofar as they offer peculiar added values to their devices, such as predictive text entry and rollers for scrolling; there are also numerous designs for hinged or sprung covers. This means that some phones do have more than the standard eighteen buttons I describe.

These differences notwithstanding, the point is that the consumer is not really confronted with a great deal of choice. Though they may see dozens of phones in a shop, and though salesmen may try and persuade them that "this one is so much better than that one", in practice, there is virtually no substantive difference between the products on offer.

Of course, the phone purchaser could look to the corner of the shop and see, somewhat hidden from view, the personal digital assistants on offer. There they will find Visors and Palm Pilots, various Psion organisers, HP Jornadas and the latest fad device, the Blackberry. In the top corner they will even find some shockingly expensive hybrids like the Nokia 9000 which offers a mix of mobile phone and PDA functionality. Yet in the UK and Europe – unlike the USA which I shall come back to – both the PDAs and the hybrids take the back seat in these retail outlets, and though they are often displayed in prominent positions in the shop window, are typically tucked away in the shop itself. Though they do sell quite well, in comparison to the sales of the "orthodox" or ordinary mobile phone, the numbers are not great. As one salesman put it to me, "They aren't where the market is at".

So confining ourselves to where the bulk of the market is, one finds that on this side of the Atlantic the action is with mobile phones. And more particularly, these look and function pretty much like each other. There is then what one might call a "sedimentation" of the form factor, i.e., a fixity in its shape and function (although this is not to ignore the aesthetic changes of fashion).

There are a number of reasons for this. One has to do with the fact that the technology inside mobile phones is becoming pretty much commoditised, with a few manufacturers providing most (though not all) of the elements necessary. This includes the software for such things as the signalling protocols and application suites, the hardware like the battery and antenna, the SIM technology and of course the cases. This is making manufacture of the devices increasingly cheap.

Another has to do with the fact that the form factor is well understood by the consumer. They know, more or less, what they are buying (or at least they think they do, as we shall see in Section 14.4). Indeed, not only this, but they have convinced themselves – with a little help from marketing and advertising agencies – of the need for these devices. There is thus at the time a vast market. So on the one hand, the consumer is happy with what they get, though they may like to change the colour and shape a little bit every now and then. On the other hand, the vigorous demand for these products means that there is little motivation for manufacturers to try anything too different. One is reminded of the advice given in technology marketing books – when the market is "in the tornado" simply ship products as quickly as you can (Moore, 1999; Moore and McKenna, 1999).

What is this telling us? Consider the convergence thesis. As numerous authors in this collection have noted (especially Sherry and Salvador in Chapter 8), it has been argued for many years that there will be a converging of mobile communications and hand-held computing. Yet in practice, if one bears in mind what we have just said, it has not yet happened. People are not buying what I called hybrids. One either buys a phone or a PDA, rarely does one buy one and the same thing. And, when it comes to the crunch, one buys a phone rather than a PDA.

This is not to say that what users are doing with their mobile devices, and occasionally also with their PDAs, is not itself interesting; as we shall note in a moment, several of the chapters in this book have highlighted just how interesting it is. But what it does mean is that there may be more resistance to the move towards new form factors and new functions and/or applications than might be being predicted. Such resistance may well be consequential as we begin to see third generation networks introduced.

Let us say a little bit more about this idea of sedimentation. What is the resistance here exactly? One source of this resistance is economic, and comes from the suppliers' end of the "value chain". Here the sedimentation of the form factor has resulted in manufacturing costs for the eighteen button phone going down considerably from, let us say, five years ago. This is why I mentioned the term commoditisation in this context earlier on. It is likely that the decrease in the cost of mobiles will continue. This will make this particular technology much cheaper than more novel technologies that have not reached a similar state of commoditisation in manufacturing – such as devices that will support continuous connectivity over third generation networks. So, notwithstanding the fact that network providers may temporarily subsidise the cost of these novel, third generation terminals in the long run there will be a substantive cost differential between the old and the new. Thus, the take-up of new devices may be slowed.

The size of the network providers' subsidising wallet aside, it might be difficult to alter consumers' preferences anyway. There may be a growing contradiction between the assumptions underscoring the current standard form and future forms. For example, take Nokia's "blue line" button for a command action. This is a standard feature of this manufacturer's current crop of devices. Pressing a blue line button (or one of them since some devices have two) supports instructions for the following commands and more: up, down, back, forward, yes, no, or select. This command entry technique is claimed by Nokia to be one of the key added values of its equipment, and has certainly had some success in terms of consumer preferences. But as we move toward more data driven services will it be so ideal?

How would one use the button for let is say, web surfing or, given an example from this book, using a WAP service? That is to say, when does the user press the blue line button for yes, no, or back? As we saw in Helyar's study in Chapter 13, users find it incredibly difficult to complete tasks given the contradiction between what they think pressing the blue line achieves, what the command actually does, and what the service (or content provider) of a WAP application wants the action to do.

Now, though one might criticise Helyar's study for putting the users in an invidious position (since they had absolutely no instructions to guide them in their task and hence were as good as blind), nonetheless, the study does make it clear that there are fundamental contradictions between the phone as "we know at the moment" and the "data phone" of the future. These contradictions are deeply related to current form factors and the intersection of that form with emerging standards for data services, such as WAP. Although the Nokia phone used in the experiment was different to the bulk of the phones on the market in that it has the blue line command button, the general thesis I am putting forward here – that there is a sedimentation or a standardisation of the form factor and that this may inhibit the future market – may still hold true.

Even if it doesn't, what we have seen in the preceding chapters is that the actual use of mobile phones is creating a host of changes – some beneficial to users, others more problematic – that in their own way may be creating other forms of resistance, or perhaps more accurately, may be leading the market to unexpected places. As Cooper noted in Chapter 2, these changes are forcing a reconsideration of what mobility means, what the mobile society may consist of, and even hitherto uncontentious distinctions between what is perceived to be public and private.

For example, mobile phones are changing our co-proximate behaviours in ways that are making the world perhaps a lonelier place than before. In this book, Townsend remarks on how urban space may be altered through mobile communications, making notions of what is an urban centre and an urban periphery almost redundant. Previous notions of space and place dissolve. One consequence of this, pointed towards in Chapter 5 but expounded elsewhere as well, is that mobile phones are changing interactions between people who are near one another. For example, when people are lost they don't ask a nearby stranger to help them, they call their partner on the mobile. When they are waiting, say at a bus stop or other public place, they don't talk to the stranger sitting beside them, they call a friend. Without wanting to say too much about the importance of small scale interactions with strangers, it would not be unreasonable to say that they are likely to be important in sustaining society. There is a sense in which they create what one might call "social connectivity". This is related to the number of people an individual may converse with on a day-by-day basis. What Townsend's argument in Chapter 5 suggests is that instead of increasing communication within society, mobile communications may in some ways reduce them. One paradox is that as social connectivity may go down, call volumes, meanwhile, will go up. The network operators will be happy but society as a whole less so. One wonders what Durkhiem, who first explored such dislocations of society through the concept of anomie, would think of this unusual technological paradox.

In other, less stark ways, the "performance" of society is changing. For example, ritual patterns of address which involve summons-answer sequences (sometimes called adjacency pairs in the literature) are not so much being rewritten with the use of mobile phones as being invoked in new and slightly nuanced ways. In Chapter 6, Murtagh notes that when a mobile phone rings in a public space, receivers of the call have to respond not only because the caller wants a response but because those around the receiver exert a pressure on them to do so. People sat beside the recipient glance in various kinds of purposeful ways to express a moral pressure on the receiver of a call to answer. Or rather, this appeared to be the case when mobile phones first came into widespread use. Today, the need to answer is treated a little differently, with more widespread recognition that "it is OK" for someone to choose to ignore or redirect a call, once they have seen the caller ID. The lesson here is that the ritual requirement of answering a summons is not being altered, it is rather that the performance of the rule is being delayed. Of most importance, the identity of the summoner determines whether the response is made immediately or at some time in the future. In this way, the ritual patterns of social relations remain, except that these patterns are being played out through time. Thus mobility here means temporal mobility, not physical.

Sherry and Salvador's chapter explored the same issues and reported that users are finding that the technology is forcing them to "work hard" at their life in a number of ways. Most particularly, there is a tension between being available any time, any place and getting things done. When someone is away from the office, for example, they have some task that they need to focus on. This could be a face to face meeting or a site visit. When someone calls them on the phone, that interrupts their focus on the local activities, distracting them from their primary goal. To solve this problem, users have to develop techniques that shield them from unnecessary interruption.

Interestingly, Sherry and Salvador go on to argue that the problems come from the connectivity technologies alone (i.e. the mobile phones) and not from the digital assistants they might also use. For these latter objects are "transport technologies": that is devices used to carry information to places where it is needed. In contrast, connectivity technologies create opportunities for others away from some particular site of action to create demand on an individual. Thus the mobile phone, though clearly a very useful tool, needs to be restricted and controlled lest its inappropriate use overly disrupts and undermines work. The view of PDAs is not all rosy, either, and Sherry and Salvador remark on how poor they are at supporting ad hoc data entry tasks, particularly note-taking. Even so, the take up of PDAs, particularly in the USA, is now becoming great, despite these problems. Perhaps it is that all they are is transport technologies, and for this, rather than the ad hoc data entry tasks, they are satisfactory. In any event there might be other reasons for the popularity of PDAs that I will come back to shortly.

In Green's chapter we find that although kids are often given mobile phones by parents who think that thereby their kid's safety is (better) assured, in practice, the kids turn the tables on their parents. Instead of the parents surveilling the kids, it is the kids who are surveilling the parents. They do this through using caller ID to ignore calls from parents as and when they see fit; by using texting to communicate beyond the monitoring of parents (and teachers and others who might have hitherto had a right to monitor). We also learn that the traditional concern of sociological inquiry with technologies of surveillance is of limited use when it comes to unpacking these aspects of mobile device usage. For the term surveillance, as used by sociologists like Foucault and those in his thrall like Dandeker, bears little resemblance to the actual practices of parents and kids themselves. Using mobile devices to resist – or more accurately avoid – parental control is only part (and indeed a lesser part) of what teenagers use mobile phones for. Hence, the suggestion that the mobile phone enables parents to "survey" their offspring is empirically off the mark; perhaps more importantly, a distraction from what is of empirical importance.

So what then are kids doing with their phones? And what then do parents think they are doing when they pay for them? These are questions I shall come back to. But one of the ways one can presage what I will consider is a finding from Gant and Keisler's chapter: that the mobile phone, constituting a technology that blurs the boundaries, is also one of those technologies that is treated in a peculiar fashion. Unlike PCs, unlike even PDAs, mobile phones become highly personal objects, treated with as much fetishism and sentiment as one can imagine. Take them away and you don't get complaints that people will be inefficient, unable to get their work done, no; what you get is anger and fear. Mobile phones are matters of the heart, not the mind (see Harper, 1996).

Let me conclude this section by considering why certain technologies seem to get taken up for reasons which are not simply economic or functional. Here I am thinking of why it is that there seems to be a difference in the acceptance and widespread use of PDAs in North America and in Europe. One recent American visitor at my research establishment commented that he was struck by the absence of people using Palm Pilots in meetings in the UK. "I couldn't do without mine," he said, "I would feel completely at a loss."

Estimates of penetration of PDAs in the American market are not entirely helpful when it to comes to explaining this. Palm themselves brag about their 70% share, but as to who or what sector of the marketplace buys either their products or PDAs more generally, there is less information in the public domain. Forrester reports that about half of US households don't have either PDAs or mobile phones, and that somewhat over a third are unlikely to ever own either (Forrester, 2001). Nevertheless, anecdotal evidence seems to suggest that for professionals, particularly mobile professionals, PDAs and Pilots especially have reached very high levels of penetration. They are, as they say, *de rigueur*.

Now if it were the case that the take up of PDAs could be accounted for solely on the functional benefits they provide, then it would also be the case that those benefits would show themselves in markets outside of North America. If sales are anything to go by, these functional benefits do not seem sufficient themselves, certainly not in Europe. Again it could be countered that the European consumer has not reached the same state of familiarity or sophistication as the North American, and when they have reached that stage these benefits will seem clear. But this would be to ignore two facts. First, Europeans have been much more willing to buy mobile devices than Americans historically so in this regard they have already had the sophistication; and second, PDAs have been available in Europe for a substantial period so any claim that the consumer is unfamiliar with them is hard to countenance.

Though it is difficult to say this with certainty, it would appear, then, that there is something else that motivates the ownership and use of PDAs in North America. One element may have to do with an emotional bond, alluded to with the quote above. This may be related to cultural systems of meaning. Unfortunately, there seems to be no literature exploring this. Hopefully this will begin to appear soon. Another element is likely to do with how the devices themselves fit within a larger information ecology (Sellen and Harper, 2001). The distances North Americans travel will be important here, as well as the time they spend away from the office. Accordingly the role of PDAs as transport devices, as Sherry and Salvador suggest, may have greater importance in the USA than on this side of the Atlantic. But this has not been explored widely and the Sherry and Salvador chapter, good though it is, reports studies of uses in both Europe and North America without highlighting any differences between them. Much more needs to be done.

14.3 Radical Form Factors

Now the network operators and the manufacturers might respond to the preceding arguments about the fixing or sedimentation of form factors by saying that it is about to change with the imminent introduction, albeit in a somewhat piecemeal way, of third generation networks. These will deliver larger data rates than current networks (but not that large, it has to be said) and when this happens, we are assured, users will find it much easier to use the devices. Video telephony will do away with the need for arcane technologies like SMS; confusions about such things as goals and purpose and urgency of the communications will be solved by "seeing the caller" and "seeing the document in question", and so on (see for example CHI 2001 Panel: Mobile Futures). What is being said here is that data rates enable new form factors, and those same form factors can enable new uses of the expanded potential of the networks. What is being thought of here is not simply how Palm Pilots might start having mobile communications functions (they already have), but also the more futuristic possible evolution of form: wearables, wireless jewellery, and so on.

As it happens, none of the chapters in this book have looked at these future forms directly, though at least two have looked at it somewhat indirectly. Churchill and Wakeford, for example, present a schema that can enable designers to determine what kind of services and devices will support "loosely" or "tightly" coupled activities. O'Hara *et al.* in Chapter 12 go further by pointing out some of the ways such things as paper documents can be "interacted" with through augmented hand-held communications devices. But before we get on to any of those possibilities, let me just wrap up some of the arguments from the prior section. For there is one aspect of user behaviour that I have not yet mentioned that might create some resistance to these more radical form factors.

Consider, when one starts presenting users with a host of different forms and interaction modes, a number of difficulties arise. The most important has to do with what are called modal shifts. Interaction designers make an especially big deal about the modal shifts that occur when users have to move between, for instance, interactive voice recognition systems (IVR) and keyboard entry for different services. This is because the user finds such shifts confusing, though in the literature just what is meant by confusion and why it manifests itself is open to debate. Be that as it may, it should be fairly clear that even the most simple services on current mobile devices require multimodal interactions and this can create confusion for users – consider many answering services which require both keyboard actions and voice. If one starts to think about more radical form factors each with their own modes of interaction, then it should be clear that users may find it ever more confusing. In other words, differences in products may lead to resistance, not take up. Users will find the equipment and the diverse ways of handling it just too complex.

In any case, the path toward the data rich future will increase the kinds of confusions users have to deal with – creating, if you like, modal shifts by default if not intention. For the fact that third generation networks are being introduced in a piecemeal way, as I said at the outset, may well mean that users are confronted with somewhat muddled devices which require a host of operations and control techniques. The devices will sometimes work on third generation networks and sometimes on intermediary networks. In each case, the services available, and hence also to some degree the way those services are interacted with, may vary. And whilst this is going on, there are also the "curios" of current design around the soft button options, such as predictive text entry, which the user will need to learn and then, presumably, forget as they are replaced with better designs in the future.

One consortium in the market place is trying to deal with these issues by creating standardisation. Though it is often thought to be driven by the need to manage the limited computing capacity of the devices themselves, Symbian – a consortium of mobile terminal and PDA manufacturers – is also working on simple or standardised solutions so that the users themselves don't have to use excessive amounts of brain power when operating third generation devices. It has been working on standards for interfaces icons, and procedures for navigation on screen. These have considerable implications for the design of hardware too. The result may be to reduce the modal shifts required in interaction. It is not at all clear, however, how successful Symbian will be in having its views taken up widely.

So, these are some of the problems one has to think about as the market place moves from what I have called the sedimented form factor we are familiar with to more radical, perhaps more innovative forms. If one forgets for the moment the problems that the user might encounter on the path and focus just on the future itself, what are we talking about?

There are numerous instances of radical forms factors reported in the literature, coming from research labs in Japan, Europe and North America. One typically finds that this research involves combinations of virtual reality headsets, haptic interfaces and so-called wearable computing. Much of it, it has to be said, is downright silly; though sometimes the ideas do seem better thought out. There is quite a lot of research on wearables, for example, and this explores how such technology can support new forms of e-commerce. There are numerous papers reporting on the idea of using a "wearable" device in the shape of a broach. This "bleeps" the wearer's unique identity to devices placed at the threshold of shops. Whenever the wearer passes that threshold, a contact is made. With this data provided, further information can then be sent to the wearer, either as a web document or even as a brochure posted through the mail.

Some of these papers explore the relationship between current and future behaviours, but most do not. This is one of the reasons why I don't want to cite too much literature here: most of the stuff on radical form factors is simply toy building. There is not of course anything wrong with this: it simply falls beyond the pale of my concern.

One exception within this book has been the previously mentioned chapter by O'Hara *et al.* (see also O'Hara *et al.*, under review). Here they report ongoing studies of mobile professionals. What they are finding is that these individuals are putting aside mixes of kit and preferring instead to use the lowly mobile phone to "do the work for them". That is to say, instead of loading up their laptops and palmtops to make sure they have all the things they need, their experiences of the difficulties of ensuring this has resulted in them using the mobile phone as a way of summoning others to do the work for them. The others are, needless to say, colleagues and secretaries back in the office.

Nonetheless, O'Hara *et al.* do discover that much of what mobile professionals want to do relates to documents and they go on to propose a way of augmenting the kinds of interactions that mobile professionals might have with documents. They outline a mixture of changes in the functionality of the mobile devices – to include scanning for example – and changes in the documents, such as through adding data glyphs. Key, though, is that there appears to be little demand for document creation and manipulation, with greater emphasis on access and distribution. In these ways, then O'Hara *et al.* look at the present to try and determine what may be useful in the future.

Interestingly, researchers at Xerox have been investigating similar possibilities. Eldridge *et al.* (1999) and Lamming *et al.* (2000) report on the design of Satchel, which supports the ad hoc use of documents. Through inventing the idea of document tokens, which are essentially the URLs for documents combined with access rights, they propose a technique whereby users can retrieve or share documents any time and any place. Tokens can be beamed between PDAs or to nearby fax machines or printers, so long as those peripherals have been "Satchel enabled". The technology relies heavily on the web, and can be implemented fairly easily as an application suite on most PDAs. Yet Xerox have had only limited success with this product. It would seem that Sherry and Salvador's view that PDAs are used as transport and not connectivity technologies holds true, even when users are offered simple solutions for bringing the two together, as would be the case with Satchel.

Irrespective of whether the Satchel application will ever take off – which seems unlikely now – the point is that there doesn't seem to be a killer application in this area. By this I mean that if radical form factors involve augmenting or transforming the shape and function of mobile devices, then if one looks at the literature there doesn't seem to be anything that has succeeded in this. The diary and time management applications associated with the Palm Pilot seem to be, at the current time any way, the limits of PDAs; SMS and audio-telephony the limits of mobile phones. Attempts to break out of the mould, as in the case of Satchel and as pointed toward by the Hewlett-Packard researchers, don't seem to be pushing the market along. This is not to say that a market will never appear; it is to say that one cannot tell from the research that looks at current practice.

14.4 User Perceptions of the Mobile Interface

When someone uses a mobile device, what they interact with is not simply a piece of hardware nor yet the software on the device itself. They are also interacting with a network provider and a service agreement. These affect what they can do as well as what they think they can do. From this view, the interface is an amalgam, and consists of much more than is normally accepted. Broadly speaking, research into this issue is well illustrated in Chapter 10 by Palen and Salzman. This reports that users "construct" a rather complex understanding of the interface. This merges and often muddles the hardware, the software (menus and display based controls) and what Palen and Salzman call the netware that a provider makes available (i.e. the basic mobile telephony service and special services such as advanced versions of call forwarding provided over the network). It also muddles up the bizware, the details of the service agreement, including calling plan, sales policies, and customer care.

In other words, and to put this in sociological terms, the mobile phone – as an amalgam of technological affordances and commercial rights – is a socially constructed object. Now, without wanting to say anything about whether such objects are "right" or "wrong" or even to imply anything about the relevance of those terms for socially constructed objects, what one can say in this instance is that this leads users to muddle up some of the all too concrete facilities in question. As Palen and Salzman report, users confuse network problems with problems with the terminal when, for instance, they wrongly think there is a lack of network coverage when in fact their phone's antenna is not very powerful. Different phones would not have this problem (though different phones may have other problems). They blame the network when they cannot use the phone outside their home area when in fact the bizware contract they have signed prohibits this. They confuse the hardware for the software when they say they cannot adjust the volume, "there being no knob to do so", when in fact, they may be able to adjust the volume through a software protocol. These are just some of the examples Palen and Salzman list.

These are of course prosaic problems, and indeed are indicative of users who have been badly briefed when they purchased the devices. Doubtless users will increase their familiarity with the technology and what it offers. Given that the Palen and Salzman research is US based, it might also be the case that some of the difficulties and muddles reflect the fact that widespread availability of mobile networks is fairly new in the US; there is even now still considerable variation in quality of service and coverage. But it seems unlikely that the muddling of the contract, the software, the terminal and the network will disappear altogether in the US or elsewhere, though it is surely the case that these issues will play out differently across cultures and countries. Consider how many different contracts and agreements that users are subject to when they buy a mobile phone in the UK at present. It is not just that there are different tariffs, but these are often related to different phones (i.e. some tariffs are available with some phones but not others). Even the most astute consumer will find the differences and specifics between various options rather daunting; how much more so for casual users who simply "want the best deal"?

The real lesson here is not simply that the current situation is complex; it is that it is likely to get worse with third generation devices and networks. After all, it is almost certain that billing will become more complex. When someone buys a phone, they will not only be buying an "internet enabled" device but also a set of services that they can use on that device. There will be bills for voice calls, for example, for continuous connections, for ad hoc services, and for downloading.

Of course it may be that the network operators will want to simplify the process. One solution would be to provide a "one charge for all" deal. As it happens this is pretty much what is available on the largest network operator in Japan. Without going into the details of the Japanese procedure, such a charge would be a standard fee per month. But this may not be what users want elsewhere, in the USA or Europe; it can make the mobile network seem expensive, especially for those who don't use the full range of services. To ensure that users can pay for what they need, then, network operators may continue to offer discrete levels of service and may also continue to subsidise terminals accordingly. Given this, similar problems to those that Palen and Salzman describe may manifest themselves in the future.

Clearly these issues – or rather sources of potential confusion – could be addressed, even if the operators continue to opt for what may appear to be rather complex arrangements over pricing, services and terminal provision. But it is likely to be difficult to do so. Current (though unpublished) studies of point of sale activities in the UK, for example, indicate that consumers walk out of a mobile phone shop knowing little about what they have bought apart from the most minimal information about monthly network rental costs. They may not have even seen the phone they have bought functioning, since phones are not charged up or connected in the shop.

Furthermore, the current organisation of sales incentives emphasises numbers of phones sold over and above anything else, such as quality of user understanding of services and contractual agreements. This may mean that there is some resistance from sales staff who will see no benefit in spending more time explaining to consumers what it is they are buying. Salesmen do better by simply flogging more phones.

Finally, post hoc customer sales support is expensive, and though operators are increasing the levels of this support as a way of reducing "churn", this avenue too will be unlikely to correct all the misapprehensions and muddles the user may have. It is simply not possible to lead the user through all the complication of using the technology over the phone or via a website or portal.

All of these matters have to do with the use of mobile communications over mobile networks. Now let us consider this in light of the future possibility of interactions between mobiles and other nearby devices. In this situation not only is the issue one of understanding the relationship between the mobile phone and the network, the bizware and so on, but the relationship between the phone and other devices nearby. Here there are additional protocols for communications that the user may need to understand. Bluetooth is obviously one such protocol, though IR has a much longer standing provenance in this area as a way of communicating locally. Whatever the standard, the point is that local interaction will add complexity to the situation. As we saw in Chapter 4, and as was explored more abstractly in Chapter 11, the actual patterns of use of objects both near and far, are remarkably subtle and complex. It requires an artful management of objects, processes and know-how; it brings the interdependencies of informational artefacts and human action; and above all it needs to be managed in a way that ensures economy of effort. Users will try and bring in new technologies to help them achieve their ends, but if they find that doing so slows down what they can do by making the process more complex or fragile, then they will turn away from it. And as Laurier notes, some of the places in which this fragility may show itself – such as while driving – may even cause risk to the user and others.

The long and short of this is that the interface to the mobile, when understood as a social construction, is more complex than the interface on current PCs. For whereas a PC basically consists of a set of applications on a virtual desktop, when that PC is connected to the web, all the user pays for is the amount of time they are on the line (notwithstanding purchases of goods). Crudely speaking, all the user has to get their head around is using the PC and paying for a phone bill. With a mobile, they have to get their head around much more. Though the devices themselves may be small, the problems that users have in understanding what they have in their hand are then much more substantial. As we move forward to a situation where new services will be made available, and then when this is combined with local or micro-level communications between nearby devices, these difficulties will increase.

These are some of the additional reasons the data rich future may be resisted over and beyond what I discussed in Section 14.2. For what I am saying is that even at present, users can find themselves confused as regards what they think they have in the hand. This inhibits the use of not just the device itself, but the amalgam of opportunities that make up the technology when understood as a whole. For mobile technology it is not just a piece of hardware nor the software; it is services that might be available remotely, communications protocols between devices locally, and contracts and associated billing and tariff processes. As users are provided with more options, they may find themselves subject to ever more confusions and misapprehensions. Attacking the problem in the ways I have suggested above does not look promising, whether it be through simplifying costs or more effectively educating the users, since there are countervailing implications in each case. So, the problems I have sketched out in this section might all be in the mind of the user, but they are all too real for that.

14.5 Interfaces to Social Process

In this last section, I want to explore what one might learn about the general relationship between the overall affordances of mobile interfaces and various patterns of social action. Put simply, I want to ask: what is known about what people are doing with mobile devices and how do these doings "interface" with the technology itself? Of course this is to stretch the meaning of the term interface somewhat, but if the reader can bear with me, I hope the reasons for doing so will become clear.

The texts in this book cover mobile professionals, people in public places, teenage and family life. Since I have said quite a bit about all of those in one way or another, I want to focus on one group in particular, teenagers and their use of SMS. The way I want to approach this is by asking whether current research suggests that SMS could be replaced by richer forms of communication enabled by new interfaces on third generation devices.

Though it might seem odd to say so, there is already a traditional view about why SMS is so successful. I say traditional since it fits into what I think of as pretty normal forms of explanation. It holds that SMS is popular because it's "quicker, cheaper and easier".

But is it? After all, if this was the case, then arguments about substitution of, let us say paper-mail with email would be easy to make. Email is cheaper, certainly quicker, and with continuous connectivity provided by, say, ADSL, easier. But as I have noted elsewhere, attempts to use these criteria almost invariably turn out to be wrong (Harper, 2001a,b; Harper *et al.*, 2001). So there must be other reasons that account for the take up of SMS.

Let's begin by unpacking these criteria. In terms of speed, most researchers (including Eldridge, 2001) report that typing on a keypad is not much of an issue for "texters". They can enter words fairly quickly – though this is not to say that they do it as fast as they might with let us say, a Qwerty keyboard. Data entry also has certain features. For example, texters don't always look at the screen when they're typing: somehow it's easy enough for them to do it while they look at other things, like watching the road ahead of them when they are driving. Texters often use only one hand. They also use many abbreviations and shortcuts, though this seems to have less to do with sustaining (or creating) a new language than it has to do with the fact that it makes communicating quicker. For the same reason, they avoid double letters.

But beyond these merely mechanical aspects of data entry, it seems to me that what makes SMS "quick enough" is that it constitutes a new communicative form or "interactional pattern". With SMS, senders don't get a reply. When one sends a text message one is really sending a kind of note that informs the recipient but is not expected to lead to a response, at least in the immediate subsequent period. That is to say, it does not lead to a series of immediately sequential turns at talk, as in a face to face conversation. The fact that there isn't a next turn has some peculiar advantages: it means, for example, that SMS is a popular way of avoiding the embarrassing next turn after a girlfriend or boyfriend has been dumped. According to Eldridge's informants, for example, it is for this reason that SMS is the number one method for "dumping your boyfriend". They don't come back and ask: Why?

Leaving aside this particular aspect of teenage life, what this new interactional pattern is about is, if you like, sending "gifts", not messages to respond to. New kinds of gifts are emerging as a result of the take up of SMS, such as sending "Goodnight" messages. These are sent with no expectation that they will be responded to. They are, in this case at least, quite literally the last word. Gifts don't always have to be wanted of course, as the example of breaking up with boyfriends makes clear.

When one begins to see this, it also becomes much easier to understand what it is that the teenagers described in Weilemann and Larsson's chapter when they exchanged the phones among themselves when sat in cafés and other hangouts. This is because the phones are the receptacles of the gifts sent to the owner, and thus to view those same gifts and treasures one has to look inside them. So it is that teenagers give their mobiles to one another so that they may look and laugh and presumably sometimes sympathise at what they find inside. Of course, one of the curious consequences of this is that it is the device as a container of gifts that is mobile more than anything else. Here too we see a need for delicate understanding of what the term mobile technology means.

As I mentioned at the end of Section 14.2, the affordances of mobile devices as a whole are, then, emotionally valued. In this book, the chapter by Gant and Kiesler (Chapter 9) in particular has explored this fact. Other research dealing with similar topics is also now beginning to appear. Some cultural specificities may impact on this with anecdotal evidence suggesting that the attitudes of US teenagers will not follow this path, since they already have objects that they invest a greater emotional attachment to before mobile phones were invented, the most obvious being the car. According to this view, Europeans need mobiles to grow up and get away from Mum and Dad, whereas in the USA, they have been using the "auto" for the same purpose since the 1950s. Whether the mobile will displace the auto in the affections of American teenagers, only time will tell.

So what are the lessons I am wanting to take from all of this evidence apart from the need for further research in certain areas? The lesson is that mobiles are being used to create a new communications pattern. One should not think of paper letters, fixed line phones, email, and SMS as competing communications technologies, so much as technologies that each afford different communicative patterns. The SMS pattern is, amongst other things, a "low data rate pattern". This may not be too good for future networks that can support large data rates. But that may be missing the point. If one recognises that SMS is a gift form of communication, the fact that it is currently needing low data rates may reflect the fact that users cannot send high data rate gifts. For example, users could be allowed to send multimedia gifts. But the important point is that such gifts would not be, let us say, video-telephony. For what is needed is say, video segments that are sent and stored on the devices themselves. This is required because these gifts need to be housed in the treasure box, the mobile itself. This means that though there may not be fully duplex trafficking of data (as there would be with full video-telephony), multimedia images are being sent over the network. But their ultimate destination and storage point is the device itself. Currently mobile technologies do not allow this. And yet this would appear to be what teenagers want, if one is to take the research of Weileman and Larsson and others seriously.

In addition, we have also learnt that the interface to current technologies doesn't seem to be a problem, at the moment anyway. Users recognise the limited affordances of the devices, and use them to frame opportunities for action. This has led them to the communication patterns I have described. Texting is deeply bound up to the fact that the technology makes texting about the best one can do; but in recognising that, those who text know that they can still achieve the ends they have in mind: friendship, companionability, even romance. At the same time, users have developed one-handed control skills. Are we then stuck with this interface, not only because of commoditisation, but because of user practices? It is to that I now turn in the conclusion.

14.6 Conclusions

These arguments suggest that mobile communications are being socially shaped. After all, the fact that users have a "social construction" of what an interface might be would seem one instance of how human perspectives are driving change rather than the technology itself. But to take this as the meaning of social shaping would be wrong and indeed misleading on two counts.

First of all, the term should not be used to mean that it is people who drive technological change, rather than technological development. When properly used the term is a label for the fact – and indeed it is a fact – that people and technology shape each other. Social shaping is then an assumption of the kinds of inquiries represented in most of the chapters in this book, rather than the output of such inquiries. At certain times people will do more of the shaping; at other times the technology. If one wants to get into the business of discovering which comes first, then one has missed the point. There is never an end point here: one can always go back a step further and discover that at the previous stage it was the opposite that may have driven change. In other words, such inquires end up like chicken and egg arguments – rather pointless. What one should try and do is explore how, in any particular historical instance, the form and balance of factors that constitutes social shaping manifests itself.

Unfortunately, although it might appear pretty straightforward to do this, in practice it is not. There is a tendency, especially in social studies of science and technology research, for researchers to "discover" that it is always social factors – people – that drive change, and not technological invention. According to this kind of research, technology doesn't matter. That is to say that though such researchers claim to be neutral, when it comes to the crunch all they ever find is the same thing: social factors are more fundamental then technical ones. This has the advantage, for sociologists at least, of placing them in the position of importance, analytically speaking, since instead of technologies and hence technologists holding the keys to change, it is social factors and hence they, the sociologists, that are experts on what the future may hold. I mention this not simply as an aside, but because those who undertake research into social shaping research are often afflicted by a closed mind about the facts in question. Instead of reporting what actually goes on, researchers in this vein tend to discover classic sociological themes: power, hierarchy, surveillance and so forth. What they don't do is report on how socio-technical processes are manifesting new properties or aspects of social organisation. As I said at the beginning, this seems a particular worry when we look at mobility.

For example, if one looks at the behaviours reported in this book, one finds that mobile communications devices are being used to support new forms of social action. One of the most referenced has to do with behaviour that crosses boundaries. One boundary is that between home and work; another closely related boundary has to do with the public and the private. Now it would be all too easy to misunderstand the evidence available on these matters. One often hears the complaint that work and associated behaviours are beginning to invade home life. Mobile technologies are helping in this. Yet if we look at what has been presented in this book, we find that the actual changes brought about by mobile communications are more subtle than that. Though it is true to say that work sometimes invades home life, in practice there are other changes that are more interesting and which make it clear that it may not be home that is being invaded, but work.

All of us will be familiar with how an individual's work may be subject to invasions or interruptions by telephone calls. For some reason, phone calls are treated as summons that have to be answered immediately. Now, with mobile phones, the phone calls are chasing people even when they leave the office. But changes in the ways these interruptions – or potential interruptions – are dealt with are showing themselves. As we have seen in Chapters 8 and 9, people are beginning to "work at" the problem of interruption so as to control it. They respond to calls by ignoring or forwarding some of them. They can do so in part because caller ID enables them to ascertain who the caller may be and they triage their response accordingly. This triaging is not only of mobile calls, but any call, as the distinction between fixed and mobile blurs. And yet the calls that are held off are not domestic or personal; these still get through. It is work ones that get pushed aside. So in this instance, not only is the social organisation of dealing with phone calls changing, but private calls are being allowed to invade work in ways that work calls are not.

Let us take another example of how home-work boundaries are being affected in surprising ways. One often hears that mobile communications are allowing the public to invade what is private. But it would appear that on balance the reverse is holding true. Consider what we saw in Chapters 5 and 6. In these chapters it was explained that people are using mobile communications to avoid participating in public life. They call home when they have some time on their hands, they don't talk to the stranger sitting nearby. On trains and other public spaces, they don't even gaze at each other in silence, as Simmel noted so long ago. Instead, they listen to each other's private lives being spoken about on the mobile phone. And it is not home that is invading the public space of the train, it is work environments where meeting times and other micro-management activities become audible to all and sundry.

In these ways, then, the private and work lives of people are invading public space. This is affecting the social rules of behaviour in public spaces; these are now being driven by the rules and etiquette of private and work life. Thus it is that mobile technology is changing society. What I have suggested in this chapter is that these and other changes have, unfortunately, been overlooked sometimes; the chapters in this book have gone some way to remedying this situation. Even so there is still a need for more investigation.

Of course one could say a great deal more about how one might explore these kinds of new phenomena. This is not the proper context for discussing all these issues surrounding such explorations. As I said at the beginning, my purpose in this chapter has been to review what the evidence presented in this book indicates; I have also wanted to stir up some of the thinking about that evidence. My theme has been to consider how studies of present practice can help guide our predictions as to what the future may hold. I have been wanting to avoid making any claims that the future is driven by either social factors or technical ones, wanting to say that there is a complex interrelationship between the social and the technical. Nevertheless, my emphasis has been, given my concern with current practice, on those social factors that have been affecting social shaping, and much less attention has been given to technological factors.

I want to conclude with a word on the relationship between SMS and user behaviour. From what I have been arguing, particularly in the last part of this chapter, it might appear that it is users who have been driving change. And indeed in some ways it is. But consider, I have suggested that the limited affordances of current mobile communications have been dealt with by users who have developed a new form of communicative practice. So what, then, came first? Texting or SMS? Historically, of course, SMS, and this provided the opportunity for texters to learn to text. But let us think about this more carefully. Is it not also the case that the new communicative practice reflects a need that was hitherto dormant for the lack of a technological vehicle for its expression? In which case, the social need came before the technology.

Such arguments, it should be clear, have no real end; more importantly they have no real purpose. It does not matter which came first. What is of importance – certainly what is important to me – is what this particular manifestation of the relationship between technology and social patterns of use tells us about where the future may lie. Texting, along with other practices I have described, gives some clue as to which direction users are taking technology. As new technology arrives on the scene, so that direction may be altered. Users may adapt their behaviour and create new usages. But in getting some idea as to what direction they are pointing toward at the moment, where the users may be in the future may not be such a difficult place to predict.

References

Eldridge M (2001) A Study of Teenagers and SMS. *Proceedings of Mobile Futures Workshop*, CHI 2001. Seattle, WA: ACM Press.

Eldridge M, Lamming M, Flynn M, Jones C and Pendlebury D (1999) Research methods used to support development of Satchel. *Proceedings of INTERACT 1999*, Edinburgh, Scotland.

Forrester (2001) Latent demand for a wireless web, reported in *Internet Surveys*, http://www.nua.ie/surveys/.

Harper R (1996) Why people do and don't wear active badges: a case study, *CSCW: An International Journal*, Vol. 4, pp. 297–318.

Harper R (2001a) *The future of Paper-mail in the digital age: an investigation into the affordances of paper-mail*, DWRC, University of Surrey.

Harper R (2001b) Paper-mail in the home of the 21st century: An analysis of the future of paper-mail and implications for the design of electronic alternatives (R Harper, V Evergeti, L Hamill and J Strain. *Proceedings of Oikos: Conference on Digital Technologies in Home Environments*, Aarhus, March.

Lamming M, Eldridge M, Flynn M, Jones C and Pendlebury D (2000) Satchel: Providing access to any document, any time, anywhere. *Transactions on Computer-Human Interaction, Special Issues entitled "Beyond the Workstation: Human Interaction with Mobile Systems"*, 7(3), pp. 322–352.

Lundin J and Persson L (2001) Mobilearn; Competence Development for Nomads. In *Proceedings of CHI 2001*, Extended abstracts, Demonstrations, pp. 7–9.

Marvin C (1988) *When Old Technologies were New: Thinking about electronic communications in the late nineteenth century*. Oxford: Oxford University Press.

Moore GA (1999) *Inside the Tornado: Marketing Strategies from Silicon Valley's Cutting Edge*. London: HarperCollins.

Moore GA and McKenna R (1999) *Crossing the chasm*, London: Harper Business.

Perry M, O'Hara K, Sellen A, Brown B, and Harper R (Under review) Dealing with Mobility: Understanding access anytime, anywhere. *ACM Transactions in Computer Human Interaction*.

Sellen A and Harper R (2001) *The Myth of the Paperless office*. Boston, MA: MIT Press.

Index

Out of print titles

Dan Diaper and Colston Sanger
CSCW in Practice
3-540-19784-2

Steve Easterbrook (ed.)
CSCW: Cooperation or Conflict?
3-450-19755-9

John H. Connolly and Ernest A. Edmonds (eds)
CSCW and Artificial Intelligence
3-540-19816-4

John H. Connolly and Lyn Pemberton (eds)
Linguistic Concepts and Methods in CSCW
3-540-19984-5

Alan Dix and Russell Beale (eds)
Remote Cooperation
3-540-76035-0

Stefan Kirn and Gregory O'Hare (eds)
Cooperative Knowledge Processing
3-540-19951-9

Mike Sharples (ed.)
Computer Supported Collaborative Writing
3-540-19782-6

Duska Rosenberg and Chris Hutchison (eds)
Design Issues in CSCW
3-540-19180-5

Peter Thomas (ed.)
CSCW Requirements and Evaluation
3-540-19963-2

Printed by Publishers' Graphics LLC